CHAINS OF CIRCLES

Pappus and Steiner Chains:
Classical Insights
and Modern Extensions

CHAINS OF CIRCLES

Pappus and Steiner Chains: Classical Insights and Modern Extensions

Ioannis M. Roussos
Professor Emeritus of Mathematics
Hamline University, USA

World Scientific

NEW JERSEY • LONDON • SINGAPORE • BEIJING • SHANGHAI • HONG KONG • TAIPEI • CHENNAI • TOKYO

Published by

World Scientific Publishing Co. Pte. Ltd.
5 Toh Tuck Link, Singapore 596224
USA office: 27 Warren Street, Suite 401-402, Hackensack, NJ 07601
UK office: 57 Shelton Street, Covent Garden, London WC2H 9HE

British Library Cataloguing-in-Publication Data
A catalogue record for this book is available from the British Library.

CHAINS OF CIRCLES
Pappus and Steiner Chains: Classical Insights and Modern Extensions

ISBN 978-981-98-2015-3 (hardcover)
ISBN 978-981-98-2016-0 (ebook for institutions)
ISBN 978-981-98-2017-7 (ebook for individuals)

For any available supplementary material, please visit
https://www.worldscientific.com/worldscibooks/10.1142/14496#t=suppl

Dedication:

To Professors that played a
very important role in my academic life:

Lila Marangou
Pavlos Georgiou
Stylianos Negrepontis
William F. Pohl
Johannes C. C. Nitsche

Acknowledgements

Many thanks to Hamline University and the Department of Mathematics.

Prologue

In the topics of the Plane Euclidean Geometry that follow, we present a complete exposition of the chains of circles which are usually classified in two groups: (1) the Pappus chains and (2) the Steiner chains. Many results on these topics are already published in the literature, since the ancient times, and a few of these results have found practical application too. But, the known results are dispersed in the bibliography and no complete and cohesive exposition in one book has been made. In this present book, as we develop a complete exposition, we research the topics further and discover some new and important results. Such new results are: the extension of the famous Archimedes formula to any chain and the complete determination and the properties of certain conic sections that are geometrical loci related to all these various chains. The reader must know the basic principles of the Euclidean plane geometry, the theory of similarities and inversions, the conics, radical axis of circles, etc., in order to study this book efficiently. But, for the sake of convenience to the reader, we have appended, in chapter 4, the most important facts that a reader may invoke while studying this interesting material. Some omitted details and shortened computations are accompanied by sufficient hints, instructions and bibliography, so that, they can be filled in by any educated reader. Also, a few new questions are suggested to anybody who would like to answer and publish them. The topics dealt here, besides inversion, make a great use of Descartes' Theorem for four mutually tangent circles (not at the same point) and the classical Stewart's Theorem of the geometry of the triangle. We also correct an error that has for some time existed in an item of the classical bibliography.

Ioannis M. Roussos
Professor Emeritus of Mathematics
Hamline University
Saint Paul, Minnesota, USA
iroussos@hamline.edu / infroussos@yahoo.com
Summer 2025

Biography

Professor Ioannis Markos Roussos was born on November 5, 1954, at the village Katapola of the island Amorgos, Greece. After the primary and secondary education he studied mathematics at the National and Kapodistrian University of Athens and received his BSc Degree (1972-1977). Then, he study graduate mathematics and computer sciences at the University of Minnesota and received his MSc and PhD degrees (1978-1986). His specialization in mathematics was in Differential Geometry and Analysis. He has taught mathematics at the University of Minnesota, University of South Alabama and Hamline University. Besides this book, he has published 17 research papers, 10 expository papers and the books: (1) *Improper Riemann Integrals* (first and second edition), and (2) *Basic Lessons on Isometries, Similarities and Inversions in the Euclidean Plane.* He has participated in meetings and has refereed papers and promotions of other professors. Other interests are classical music, history, international relations and travelling. Since the Fall of 2022, he is Professor Emeritus at Hamline University, Saint Paul, Minnesota, U S A.

Contents

Chapter 1

Introduction

In the plane Euclidian geometry, we encounter two kinds of classical chains of circles. The **Pappus**[1] **chains** and the **Steiner**[2] **chains**. In the chapters that follow, we are going to present known results in order to generalize them, and investigate and prove new results, on either kind of chains. Some omitted details on proofs and calculations are provided with enough information and instructive figures, so that, a familiar knowledgable reader can figure out and fill in. The different colors used in the figures help understand and decipher what we want to prove. We also provide sufficient bibliography.

In general, a **chain of circles** is a sequence of circles tangent to two given circles and each circle of this sequence (except the first) is tangent to the previous circle of the sequence. The two given circles are called the **base circles of the chain** and the circles of the sequence are **the circles of the chain**. *To avoid trivial cases, no three successive circles of the chain or two circles of the chain and one or both base circles are tangent to each other at the same point.* See **Figures 1.1 and 1.2** below, and you can draw other cases yourselves too. We do not study such chains, for they do not expose any interesting properties.

Some chains contain finitely many different circles, $n > 1$ let us say, and so we call them **finite chains**. This happens when the nth circle of the chain is tangent to the first circle. But then, we can continue along the same n circles (already constructed) as many times as we wish and so we may consider these chains as **infinite periodic chains**, or simply **periodic chains**, or **closed chains**. (These three terms are used interchangeably.) Otherwise, a chain consists of infinitely many different circles, and so, we

[1]Pappus of Alexandria, Greek mathematician, c.290-c.350.

[2]Jakob Steiner, Swiss mathematician, great geometer, 1796-1863.

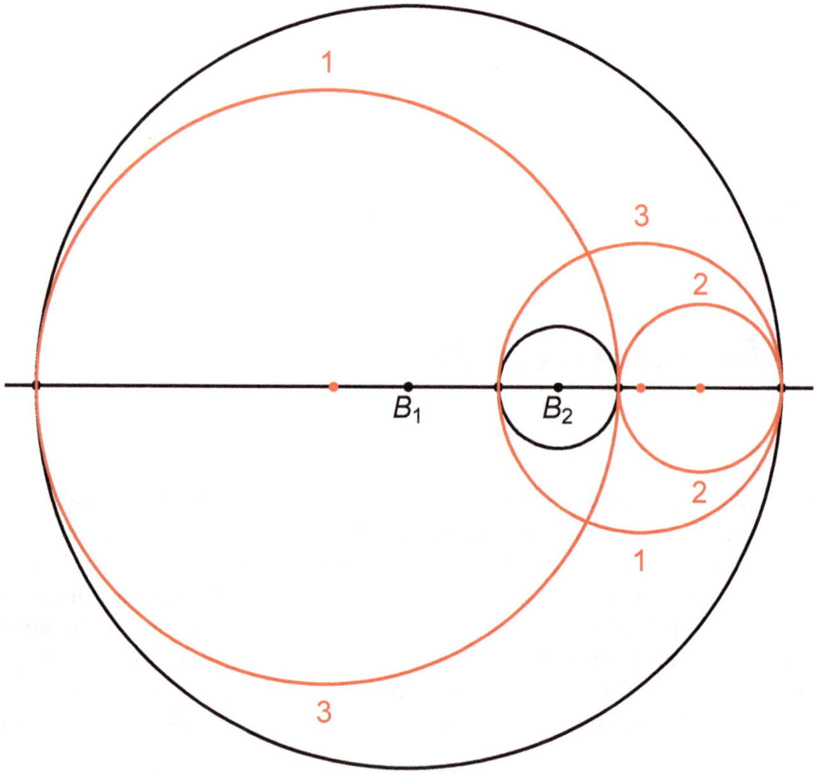

Figure 1.1: Two Trivial Chains $\{1, 2, 3\}$ with Base Circles (B_1) and (B_2), in black color.

call it **infinite chain.**

In the context of the chains of circles, the straight lines can be considered as circles. A straight line considered as a circle has infinite radius and its center is the infinity point corresponding to the direction perpendicular to the straight line. We also keep in mind that: *two different straight lines, considered as circles, are tangent iff they are parallel.*

In the next two chapters, we use many results from the general Euclidean plane geometry and the theory of homotheties and inversions. For the convenience of the reader, we have listed these results and their proofs in **Chapter 4**, which is considered to be an appendix to the previous chapters. The theorems and results listed there are also important in the study of the Euclidean plane Geometry, in general.

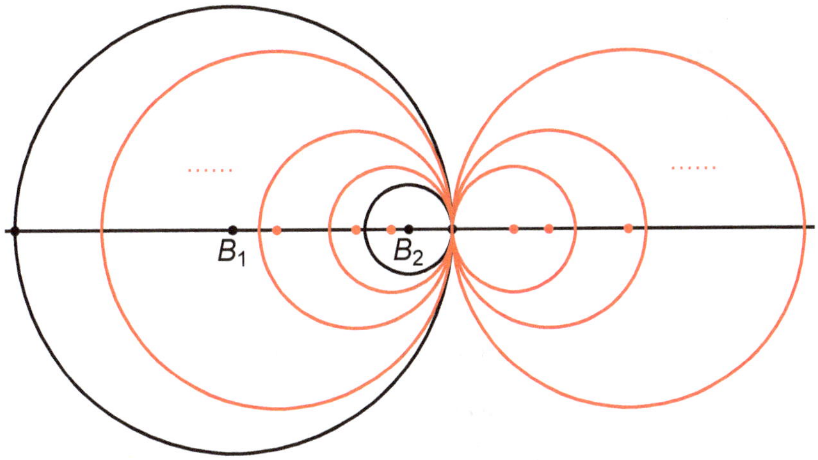

Figure 1.2: A Trivial Chain with Base Circles (B_1) and (B_2).

Some History and Remarks

The two kinds of chains of circles – Pappus and Steiner – that we study and develop in the material that follows, are two very interesting subjects of the plane Euclidian Geometry that pertains to geometrical constructions and loci. Apart from many interesting, beautiful, and intriguing mathematical properties, they also have a few nice practical applications.

The **Pappus chains** are named after Pappus of Alexandria, a Greek mathematician who lived in the 4th century CE and studied these chains. Originally, such a chain is a sequence of successively tangent circles nested within an arbelo. **Arbelo** is the figure formed by three pairwise tangent semicircles with a common diameter, two of which are inscribed within the larger one. The classical Pappus chain considers the two largest semicircles as the base circles, starts with the smallest semicircle within the arbelo and then each new circle is tangent to the two larger semicircles and the previous circle in the chain. As a result, this is an infinite chain of circles. The Pappus chain has a lot of properties and also demonstrates some remarkable properties of the arbelos. It is a classic example of a geometric construction with fascinating properties.

The arbelos were introduced and studied by the Greek mathematician and engineer Archimedes in the 3rd century BCE and discovered many of their properties. Pappus's work on chains and arbelos, in general, is documented in his mathematical collective works called **Synagogue** or **Collection**. There, we find a formula for the radius of the first circle of the chain, called **Archimedes formula**. He also included a nice theorem that he called **Ancient Theorem**. (These topics are included in this book. See also Jonson 2007, Pedoe 1970, Roussos 2022, etc.)

The Swiss Mathematician Jacob Steiner (a great geometer), in the 1800's, found a simpler proof for the Pappus Ancient Theorem using the transformation of inversion. Inversion is a very powerful tool in plane Geometry which we use in this book a lot. Using it, we discover the formula for the radius of any circle of the chain.

In the context of the topics presented, we consider a straight line as a circle of infinite radius (and center the point at infinity in the direction perpendicular to the straight line). Two straight lines are tangent if and only if they are parallel. Hence, we expand the concept of the classical Pappus chain to any chain of successively tangent circles which are tangent to two internally or externally base circles and /or straight line(s). So, the first semicircle of the classical Pappus chain is here taken as the first whole circle of the chain and the chain approached the point of tangency in both directions. Then, we expand the Archimedes formula to these cases and we also prove many interesting old and new results and geometrical loci. We extend the study of such chains when the base circle(s) and / or straight line(s) are intersecting and we find some analogous interesting results. Also, in a arbelo there are three points of tangency. Therefore, a soon as we construct the circle tangent to the three semicircles (Archimedes's circle), then we can construct and examine three different chains, each of which approaches one of the points of tangency. That is, in an arbelo, we can have three different chains sharing the circle in common.

The Pappus chains and their properties continue to be studied and explored in modern mathematics. They are connected to other areas of plane geometry and related geometric concepts. A great connection and fairly recent subject of study and research is the Steiner Chains.

A **Steiner chain**, named after the mathematician Jacob Steiner, is a chain of successively tangent circles and tangent to two fixed non-intersecting base circles. Originally, one base circle is inside the other one. In this context, **Steiner's porism**, is a great result that states: "*If a Steiner chain of n circles exists, then an infinite number of such chains with n circles also exist, no matter where we place the first circle of the chain*". Steiner also discovered the concept of the conjugate chain of a given chain and proved an amazing result between a chain and its conjugate, that he called the most beautiful theorem of plane geometry. (These are presented in this book. See also Jonson 2007, Pedoe 1970, Roussos 2022, etc.) We explain this great result extensively in this book and illustrate it with several important colorful figures for helping the readers to understand. In the material that follows, we extend the concept of the Steiner Chain to chains between any two non-tangent and non-intersecting circle(s) and / or straight line(s), and we prove many old and new results. The tool of inversion is very important in this work. We also derive a formula for the radii of the circles of the chain, analogous to Archimedes' formula.

The Pappus and Steiner chains, apart from their intrinsic beauty and mathematical interest, are also studied for their connection to other mathematical areas and applications in other fields. These include: geometry (Euclidian and Hyperbolic), Descartes four kissing circle theorem, conic sections, the contact problem of Apollonius, complex analysis, Möbius[3] transformations, topology, calculus, and mechanical constructions of ball bearings rolling between circles, and making jewelry and other ornamental products.

The material of this book is confined on the Euclidean plane. There are some studies and analogs of the Steiner chains in the three dimensional solid Geometry. An amazing fact, in this regard, is the **Soddy's[4] Hexlet** (six-bead-necklace). That is: "*Given three fixed spheres pairwise tangent externally, (leaving a hole among them), there are always six successively externally tangent spheres which are tangent to the three given spheres and (going through the hole) forming a closing chain. These six spheres surround one (any one of the three that we can choose, and so we can have three hexlets) of the given spheres*". [See Ogilvy 1969, where we find more cases of hexlets. Also, Soddy constructed a hexlet in which the radii of the nine (three plus six) spheres involved are rational numbers!]

However, there are still several open questions and areas of research for the interested mathematician to explore. These include:
(a) Possible extensions and / or improvement of the results exposed here.
(b) Extension and adjustment of results to solid geometry.
(c) Study of these chains in the hyperbolic plane and even hyperbolic space.
I suppose that many mathematicians will be motivated by this old and new subject, research the above areas and publish new beautiful results.

By the way, Pappus is also famous for the Pappus Theorem (in Euclidian and / or projective Geometry), generalized by Pascal[5]. Archimedes (280–c. 220 BCE) is well know for many achievements in mathematics and physics. A great achievement is measuring the circle and the sphere, and showing the way to estimate the number π, and his work in hydrostatics and physics. Steiner, besides his work on the chains, is well known for many results in classical geometry, e.g., the Steiner point, etc. He also gave a new proof of

[3] August Ferdinand Möbius, German mathematician, 17 November 1790 – 26 September 1868.

[4] Frederick R. Soddy, British radiochemist, 2 September 1877 - 22 September 1956.

[5] Blaise Pascal, French mathematician and engineer, June 19, 1623 - August 19, 1662.

the existence of the point of Fermat[6] - Toricelli[7] and used it for constructing minimal graphs. This point is the point that minimizes the sum of the distances from a point in the plane to the three vertices of a given triangle (e.g., see Roussos 2012).

[6] Pierre de Fermat, French mathematician and judge, August 17, 1601 - January 12, 1665.

[7] Evangelista Torricelli, Italian mathematician and physicist, October 15, 1608 - October 25, 1647.

Chapter 2

Pappus Chains

A chain of circles is called a **Pappus chain** (P. C. for short), if its two base circles: (1) are tangent internally, (2) are tangent externally, (3) intersect at two points. Then, between the base circles, we can construct chains that are infinite, and tend to the points of tangency or intersection. (See the figures of this chapter.) The Pappus chains have many interesting properties.

2.1 Classical Pappus Chains

By *Classical Pappus Chain* we mean the chain whose two base circles are tangent internally. Such a chain is called **special classical Pappus chain**, if one of its circle has center on the center line of the two base circles. Usually this circle is considered to be the first circle of the chain. (See **Figures 2.1, 2.2**, and **2.3**.) If no circle of the chain has center on the center line of the two base circles, then we call this chain **general classical Pappus chain**. (See **Figures 2.6** and **2.7**.)

When we study the special classical Pappus chains, we consider half of them (as in **Figures 2.1** and **2.3**), since the other half has the same properties as the first one, by symmetry. A well known theorem that Pappus called it ancient theorem and concerns a special classical Pappus chain, is the following.

Theorem 2.1.1 (Pappus' Ancient Theorem.) *Consider three collinear points F, G and H with G between F and H. Let $\mathcal{C}$, $\mathcal{C}'$ and $\mathcal{K}_0$ be three semicircles with diameters FH, FG and FH, respectively and lying in the same half-plane determined by the line FH. We consider a sequence of circles $\mathcal{K}_n$, $n = 1, 2, 3, \ldots$, such that all of them touch $\mathcal{C}$ and $\mathcal{C}'$ and each $\mathcal{K}_n$ touches the circles $\mathcal{K}_{n-1}$ and $\mathcal{K}_{n+1}$, $n = 1, 2, 3, \ldots$. ($\mathcal{K}_1$ touches the* *semicircle $\mathcal{K}_0$ and the circle $\mathcal{K}_2$, and so on.) We let K_n be the center of $\mathcal{K}_n$, for $n = 0, 1, 2, 3, \ldots$. Let r_n be the radius of the circle $\mathcal{K}_n$ and h_n the distance of its center from the line FH. Then,*

$$h_n = 2nr_n, \quad \textit{for} \quad n = 0,\ 1,\ 2,\ 3, \ldots .$$

The chain of this Theorem is depicted in **Figure 2.1**. The proof can be found in several items of the bibliography (see, e.g., Pedoe 1970, Roussos 2022, Yaglom 2009, etc.) But, for the convenience of the reader, we present it next.

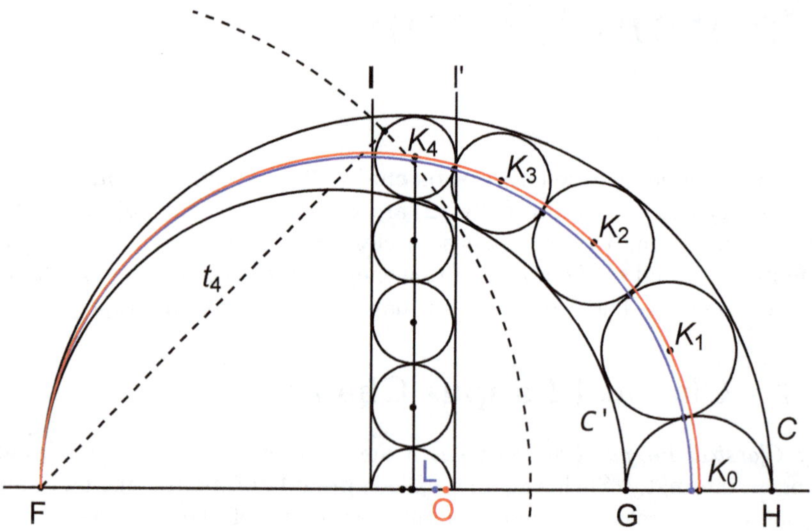

Figure 2.1: Pappus Ancient Theorem

Proof For $n = 0$ the result is valid as $0 = 0$. Next, for each $n = 1, 2, 3, \ldots$, we let t_n be the length of the segment from the point F until the point of tangency with the circle $\mathcal{K}_n$. (In **Figure 2.1**, $n = 4$.) We now use the inversion with center F and power t_n^2. So, the radius of the circle of the inversion is t_n.

Then, $\mathcal{K}_n$ inverts to itself, since, by the choice of the inversion, it is orthogonal to the circle of inversion. So, as a set of points, $\mathcal{K}_n$ is invariant by the inversion. The (semi-)circles $\mathcal{C}$ and $\mathcal{C}'$ invert to two straight (half-) lines l and l', since they pass through the center of inversion. Then:
(1) The line FH is invariant, since it passes through the center of inversion the point F.
(2) $l \parallel l'$, since the (semi-)circles $\mathcal{C}$ and $\mathcal{C}'$ are tangent to each other at F,

the center of the inversion.
(3) $l \perp FH$ and $l' \perp FH$, since FH passes through the centers of $\mathcal{C}$ and $\mathcal{C}'$ and so is a diameter line for both (semi-)circles and so FH is perpendicular to both (semi-)circles.
(4) l and l' are tangent to the invariant circle $\mathcal{K}_n$, since $\mathcal{C}$ and $\mathcal{C}'$ are tangent to it.

The semicircle $\mathcal{K}_0$ inverts to a semicircle touching the lines l and l' and having diameter on FH. The circles $\mathcal{K}_i$, $i = 1, 2, 3, \ldots, n-1$ invert to circles inside the strip between l and l' touching the two lines and each other successively and in the same order as the original circles. So, all the inverted circles are equal to $\mathcal{K}_n$ and all of their centers are on the mid-parallel line of l and l' (which is also perpendicular to FH). Therefore, the distance from the center of the invariant circle $\mathcal{K}_n$ to the line FH is

$$h_n = 2nr_n, \quad \text{for} \quad n = 0,\ 1,\ 2,\ 3,\ \ldots .$$

(In **Figure 2.1**, $h_4 = 2 \cdot 4 \cdot r_4 = 8r_4$.)

■

The shape between the three semicircles $\mathcal{C}$, $\mathcal{C}'$ and $\mathcal{K}_0$, with diameters FH, FG and GH Archimedes[1] called it **arbelo**. (Arbelo = the knife of the shoemaker, in ancient Greek.)

The chain depicted in **Figure 2.2** below can be called **complete classical Pappus chain**. Notice that, a circle of this chain has center on the center line of the two base circle and the circles. Its index is conveniently put to be zero. The lower half of the chain is the reflections of the upper half in the center line of the two base circles. The Pappus' chains are infinite and their circles tend to the points of tangency of the base circles, in either direction (upper and lower). For such a chain we have the following theorem.

Theorem 2.1.2 *The centers K_n of the circles $\mathcal{K}_n$, and K_n' of $\mathcal{K}_n'$, the reflected circles in the center line FH, for $n = 0, 1, 2, 3, \ldots,$ lie on an ellipse whose foci are the centers U and V of the boundary circles, and so its center is the midpoint O of the foci segment UV, and the major axis of the ellipse is FK_0. (So this ellipse is completely determined.)*

[1]Archimedes of Syracuse Sicily (today in Italy). Greek mathematician, 287-212 BCE. He is considered by the great majority of mathematicians to be the greatest mathematician of all times. He is the first who conceived the process of integration by which he found the area of the circle of radius r to be $A = \pi r^2$, the length of its circumference $c = 2\pi r$ and also $\int_{-1}^{1} x^2 dx = \frac{2}{3}$. He also found the first approximation of π, that is, π was less than $3 + \frac{1}{7} = \frac{22}{7}$ but greater than $3 + \frac{10}{71} = \frac{223}{71}$. In the decimal notation we use today, this value is between 3.1429 and 3.1408. That is pretty close to the approximate value usually used 3.1416.

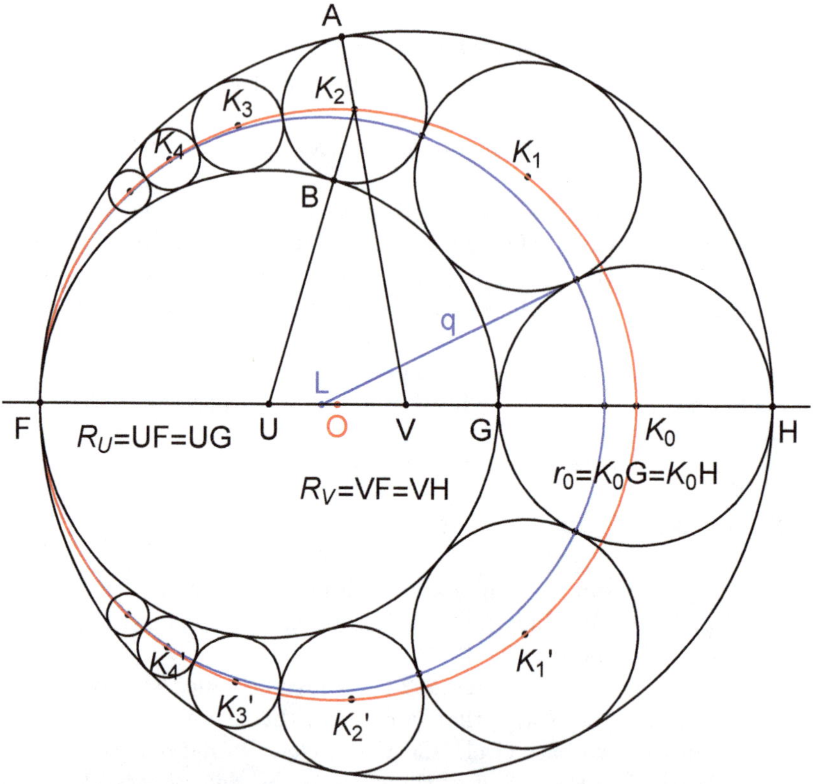

Figure 2.2: Complete Special Pappus Chain

Proof For any circle $\mathcal{K}_n$ (or $\mathcal{K}'_n$), (in **Figure 2.2**, we have chosen $\mathcal{K}_2$), we have

$$K_nU + K_nV = (AU + UK_2) + (VB - K_2V) = AU + VB,$$

the sum of the radii of the two given boundary circles, which is fixed. Therefore, according to the geometrical definition of an ellipse, the center K_n is on the ellipse with foci U and V. So, the center O is the midpoint of the foci segment UV and the major axis has length $AU + VB$, which is easily seen to be equal to FK_0. Thus, this ellipse is completely determined. The center O satisfies

$$FO = \frac{FV + FU}{2} = \frac{R_V + R_U}{2} = \frac{FK_0}{2}. \qquad (R_V = R_U + r_0.)$$

■

Remark: We easily see that the same proof holds even if the center of the circle is not on the center line FH.

In the next Theorem we prove that the points of tangency of the circles of a complete classical Pappus chain lie in the circumference of a circle for which we determine its center and radius. We also derive the famous Archimedes' formula. We refer to **Figure 2.3**.

Theorem 2.1.3 *We put*

$$r = \frac{FG}{FH} = \frac{2R_U}{2R_V} = \frac{R_U}{R_V},$$

the ratio of the small radius divided by the big radius (so, $0 < r < 1$) of the two given boundary base circles. Then, for $n = 0,\ 1,\ 2,\ 3,\ \dots$, we have:

(a) The radius of the circle $\mathcal{K}_n$ and $\mathcal{K}'_n$ is given by the ***Archimedes' formula***

$$r_n = R_V \cdot \frac{r(1-r)}{n^2(1-r)^2 + r} = \frac{FH}{2} \cdot \frac{r(1-r)}{n^2(1-r)^2 + r}. \tag{2.1}$$

$$2R_V = FH = 2(R_U + r_0).$$

(b) The projection P_n of the center K_n of the circle $\mathcal{K}_n$ (and K'_n of $\mathcal{K}'_n$) on the center line FH, satisfies

$$FP_n = R_V \cdot \frac{r(1+r)}{n^2(1-r)^2 + r} = \frac{FH}{2} \cdot \frac{r(1+r)}{n^2(1-r)^2 + r}.$$

(c) In view of $h_n = 2nr_n$, (proven previously in ***Theorem 2.1.1*** *and will be proved again below), if we let as origin $F = (0,0)$ and $K_n = (x_n, y_n)$, and so $K'_n = (x_n, -y_n)$, then*

$$K_n = (x_n, y_n) = 2R_V \cdot \left\{ \frac{r(1+r)}{2[n^2(1-r)^2 + r]}, \frac{nr(1-r)}{n^2(1-r)^2 + r} \right\},$$

$$K'_n = (x_n, -y_n) = 2R_V \cdot \left\{ \frac{r(1+r)}{2[n^2(1-r)^2 + r]}, \frac{-nr(1-r)}{n^2(1-r)^2 + r} \right\}.$$

(d) The points of tangency of the inscribed circles in a complete classical Pappus chain lie in the circumference of a circle (L, q) with diameter FM on the center line FH and

$$FM = \frac{4rR_V}{1+r}.$$

So, its center L satisfies

$$FL = \frac{2rR_V}{1+r}$$

and its radius is

$$q = \frac{2rR_V}{1+r} = \frac{2R_U R_V}{R_U + R_V} = \frac{2}{\frac{1}{R_U} + \frac{1}{R_V}}.$$

Also, this circle (L, q) intersects the circles of the chain orthogonally. Therefore, all common tangents of all pairs of successive circles of such a Pappus chain at the common points of tangency are radial lines of the circle (L, q) and so concurrent at the center L.

Proof As in **Theorem 2.1.1**, we are going to use an inversion with center F so that the straight line FH is invariant, the semicircle $\mathcal{C}$ with diameter FH and the semicircle $\mathcal{C}'$ with diameter FG are mapped to two straight lines perpendicular to FH, and therefore these two lines are parallel. (See **Figure 2.3**.) Any power $c > 0$ can be used, but for convenience we take

$$c = FH^2 = (2R_V)^2 = 4R_V^2.$$

That is, we will use the inversion $I_{[F,(2R_V)^2]}$.

Under this inversion, the point H is fixed and the semicircle $\mathcal{C}$ is mapped to the line perpendicular to FH at H. Let G' be the inverse point of G. G' is on the line FH and, with the choice of c, is outside the segment FH on the side of H that F does not lie. Then, the semicircle $\mathcal{C}'$ is mapped to the line perpendicular to FH at G'. Any two successive circles of this Pappus chain are mapped to circles externally tangent, and all these image circles are tangent to these parallel lines and therefore they are equal. (See **Figure 2.3**.)

We have that

$$FG \cdot FG' = 4R_V^2.$$

Hence,

$$2R_U \cdot FG' = 4R_V^2$$

and so

$$FG' = \frac{2}{r} R_V.$$

Therefore, the diameter of the equal image circles of the chain is

$$HG' = FG' - FH = \frac{2}{r} R_V - 2R_V = 2 \cdot \frac{1-r}{r} \cdot R_V.$$

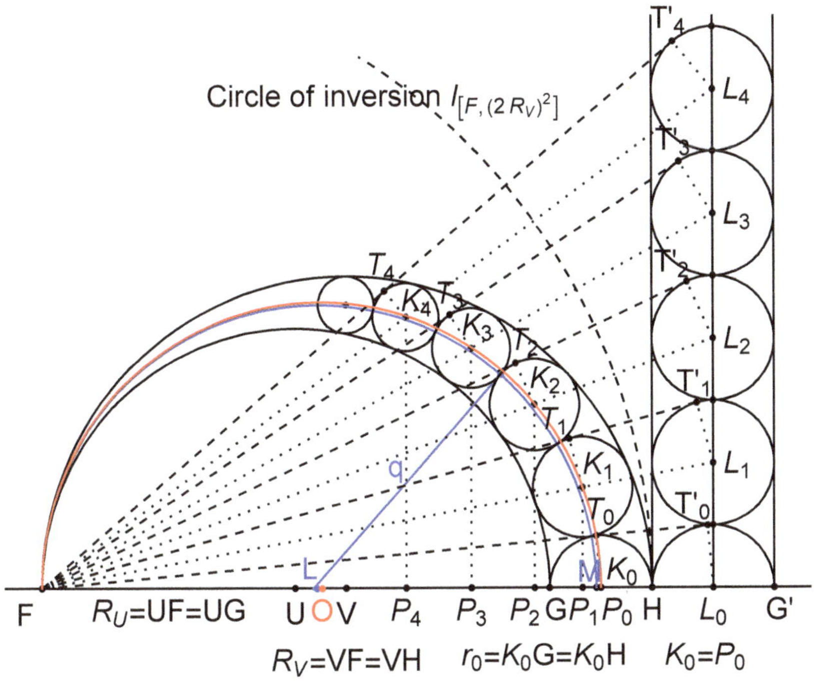

Figure 2.3: Special Pappus Chain in an Arbelo.

So, the radius R of all equal image circles of the chain is

$$HL_0 = L_0G' = \frac{1}{2}HG',$$

and so

$$R = \frac{1-r}{r} \cdot R_V.$$

Let for any $n = 0,\ 1,\ 2,\ 3,\ 4\ \ldots$, the circle of the chain $C(K_n, r_n)$ be mapped to the circle $C(L_n, R)$. Then, the distance of the center L_n from HG' is

$$L_0L_n = 2nR$$

and

$$FL_0 = FH + HL_0 = 2R_V + R.$$

So,

$$FL_n^2 = FL_0^2 + L_0L_n^2 = (2R_V + R)^2 + (2nR)^2.$$

Therefore, the power of F with respect to $C(L_n, R)$ is

$$p_n = FL_n^2 - R^2 = (2R_V + R)^2 + 4n^2R^2 - R^2 = 4R_V^2 + 4R_VR + 4n^2R^2.$$

By the fact that a circle and its inverse are homothetic, the circle $C(K_n, r_n)$ is the image of $C(L_n, R)$ under the homothety $H_{[F, \frac{c}{p_n}]}$. That is, the center of the homothety is F and using the above formula for R and

$$r = \frac{R_U}{R_V},$$

after we simplify, we find that the homothetic ratio is

$$\lambda_n = \frac{4R_V^2}{4R_V^2 + 4R_VR + 4n^2R^2} = \frac{r^2}{r + n^2(1-r)^2}.$$

So, the radius of $C(K_n, r_n)$ is

$$r_n = \lambda_n \cdot R = \frac{r^2}{r + n^2(1-r)^2} \frac{1-r}{r} R_V = \frac{r(1-r)}{r + n^2(1-r)^2} \cdot R_V.$$

The distance of the center K_n from the line FH (as in **Theorem 2.1.1**), is

$$h_n = \lambda_n \cdot L_0L_n = \lambda_n(2nR) = \frac{2nr(1-r)}{r + n^2(1-r)^2} R_V = 2n \cdot r_n.$$

The projection of FK_n on FH is

$$FP_n = \lambda_n \cdot FL_0 = \lambda_n(2R_V + R) = \frac{r(1+r)}{r + n^2(1-r)^2} R_V = \frac{1+r}{1-r} \cdot r_n.$$

The points of tangency of the circles of such a Pappus chain must lie in the circumference of a circle which is the inverse of the line L_0L_n. This is a circle with diameter FM on the center line FH, where M is the inverse point of L_0. Then we compute that

$$FM = \frac{4rR_V}{1+r},$$

etc.

Finally, the remaining claims, stated in the Theorem, follow from these results immediately.

■

Remarks: (a) We can use the formulae provided by this theorem to draw accurate Pappus chains.

(b) The radius formula is immediately correct for $n = 0$. Plugging in $n = 0$ and $r = \dfrac{FG}{FH}$, we find: $\quad r_0 = K_0G = K_0H = \dfrac{GH}{2} = R_V - R_U$.

(c) The centers of the circles of the chain lie on an ellipse with foci the centers of the two base circles. It is easily seen that the sum of the distances of any such center from the centers of the two base circles is equal to $R_U + R_V$. So, the center of this ellipse is the midpoint of the of the center segment of the two base circles and the major axis is on the line FH and has length $R_U + R_V$.

(d) We can compute r_1 using the well known **Stewart's[2] Theorem, 4.1.2.** (See the bibliography, for example, Roussos 2022, Theorem 6.5.3. See also **Chapter 4, Theorem 4.1.2.**) For this computation, we examine **Figure 2.4.**

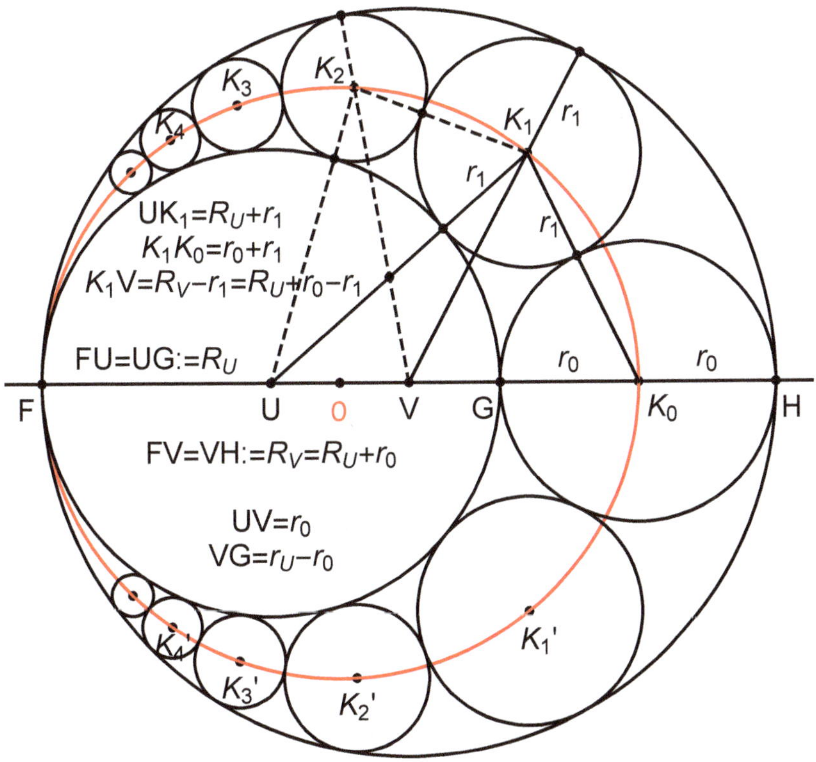

Figure 2.4: Pappus chain and calculation of r_1

[2]Matthew Stewart, Scottish mathematician, 1717-1785.

Examining this figure and checking the relations stated in it, we apply **Stewart's Theorem** to the triangle UK_0K_1 with all segments positive, to get

$$K_1U^2 \cdot VK_0 + K_1K_0^2 \cdot UV = K_1V^2 \cdot UK_0 + VK_0 \cdot UV \cdot UK_0.$$

We substitute in all segments in terms of the radii and solve for r_1. The solution is

$$r_1 = \frac{R_U \cdot r_0 \cdot (R_U + r_0)}{R_U^2 + R_U \cdot r_0 + r_0^2} = \frac{(R_U + r_0) \cdot R_U \cdot r_0}{r_0^2 + R_U \cdot r_0 + R_U^2},$$

which is the correct formula. (Check that it agrees with the formula found above, with $n = 1$. Can you find r_2 in this way?)

Another way to find r_1 is to use **Descartes Theorem**[3], **4.6.1**, **equation 4.7**, or **4.8**, with: $\frac{1}{r_0}$, $\frac{1}{R_U}$, $\frac{1}{r_1}$, $\frac{-1}{R_U + r_0}$ (do not forget that $R_V = R_U + r_0$ and we change its sign for the different kind of tangency, from external to internal) and solve for r_1. (Check it out.)

So, we see that in this chain, besides inversion, we have two ways that do not depend on inversion for computing r_1. This provides another way for computing the homothetic ratio regarding the first circle of the chain and its inverse image.

(e) The general formula for the radius of the circle $C[K_n, r_n]$, as in **Figure 2.2**, can also be written in the convenient form

$$r_n = \frac{(R_U + r_0) \cdot R_U \cdot r_0}{n^2 r_0^2 + R_U \cdot r_0 + R_U^2}, \quad \forall \quad n = 0,\ 1,\ 2,\ 3,\ \ldots .$$

Also, with $F = (0, 0)$, for $n = 0,\ 1,\ 2,\ 3,\ \ldots$, we have

$$x_n = \frac{R_U \cdot (R_U + r_0) \cdot (2R_U + r_0)}{n^2 r_0^2 + r_0 R_U + R_U^2}, \qquad y_n = \frac{2n \cdot (R_U + r_0) \cdot R_U \cdot r_0}{n^2 r_0^2 + r_0 R_U + R_U^2}.$$

(f) If we switch the roles of the circles $C[K_U, R_U]$ and $C[K_0, r_0]$, we find a new Pappus chain, as in **Figure 2.5**. [Compare this chain with the chains in **Figures 2.2** and **2.4**. Here, the circle (L, q) is indicated by (C, R).]

Then, in the previous formulae we interchange R_U and r_0 and we find that the radii of the tangent circles of the chain are given by the formula:

$$r_n = \frac{(r_0 + R_U) \cdot r_0 \cdot R_U}{n^2 R_U^2 + R_U \cdot r_0 + r_0^2}, \quad \forall \quad n = 1,\ 2,\ 3,\ \ldots .$$

[3] René Descartes, French philosopher and mathematician, March 31, 1596 - February 11, 1650.

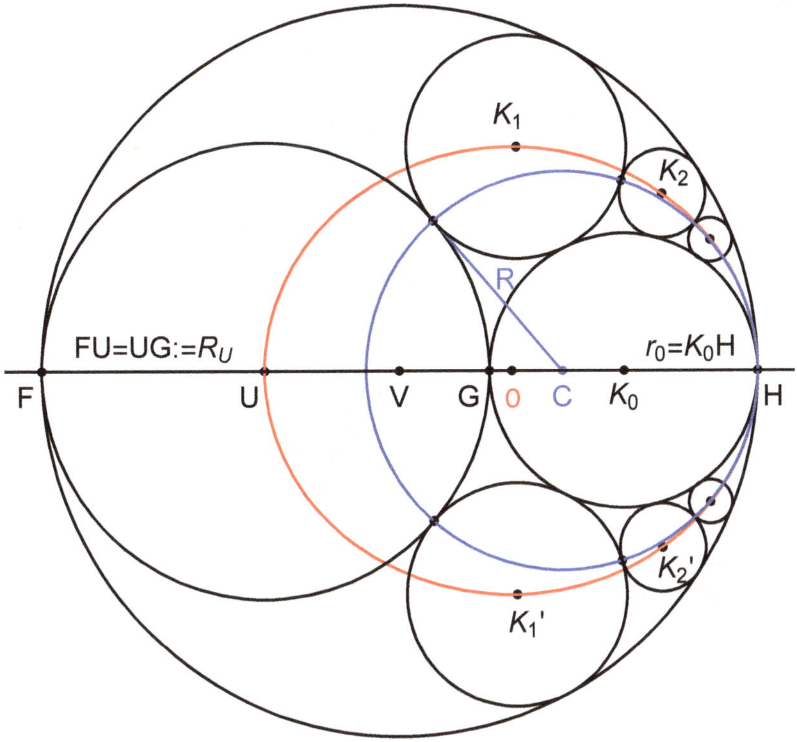

Figure 2.5: The other Pappus chain in an Arbelo

For $n = 0$, we find the radius r_U, in this case. For $n = 1$, we find again

$$r_1 = \frac{(r_0 + R_U) \cdot r_0 \cdot R_U}{R_U^2 + R_U \cdot r_0 + r_0^2} = \frac{(R_U + r_0) \cdot R_U \cdot r_0}{r_0^2 + R_U \cdot r_0 + R_U^2},$$

the same as before. In both cases the circles $C[K_1, r_1]$ and $C[K_1', r_1]$ are the same. (Expected, since they are tangent to the same three given original circles.)

{Work out and state the formulae for the centers (x_n, y_n) of the tangent circles $C[K_n, r_n]$, in this case, which are the analogous to the formulae of the original case. Taken as origin this time the point $H = (2R_U + 2r_0, 0)$, determine their positions relative to point analogous to the formulae in **Theorem 2.1.3, (c)**.}

It would be interesting to study Pappus's chains in which the initial circle of the chain K_0 is in any general position and not just on the diameter GH. We examine this in the next section.

2.2 General Pappus Chains

All of the above results exhaust the main results of the special classical Pappus chains (that is, the chain in which the first circle and the base

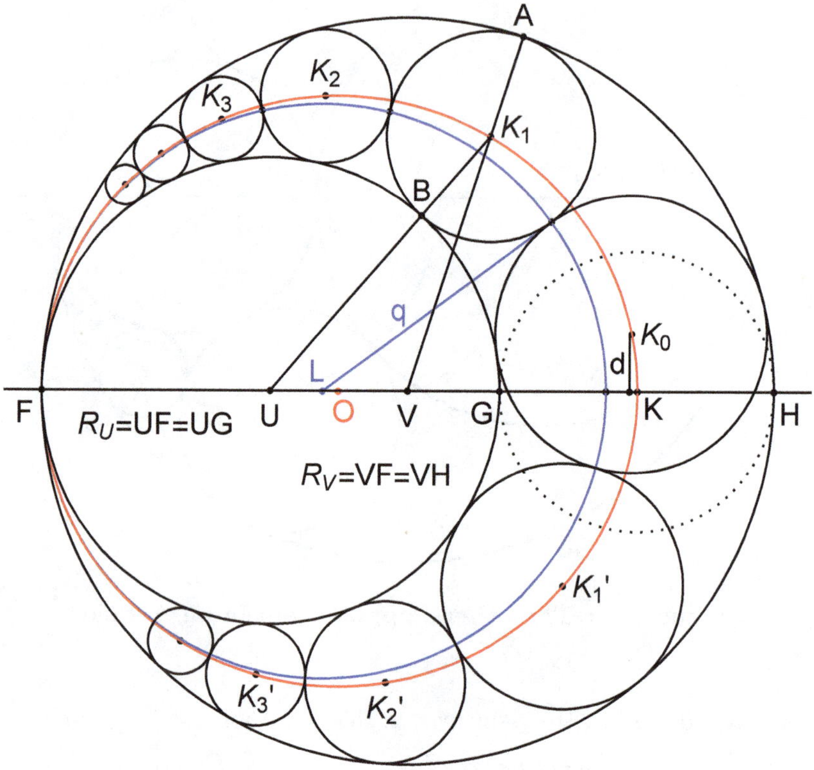

Figure 2.6: Complete General Pappus Chain

circles have collinear centers). So now, it would be interesting to investigate analogous results for any Pappus chain between two internally tangent base circles, in which no circle of the chain has center collinear with the centers of the base circles. See **Figures 2.6** and **2.7**. We can call such a chain **General Classical Pappus Chain** and / or **Complete General Classical Pappus Chain**. So, we are going next to derive the new results analogous to the results of **Theorem 2.1.3**.

Notice that in these figures, the circle that intersects the center line of the base circles does not have its center on this line. (We can say that in a general Pappus chain, the initial circle can be place anywhere between the two base circle and then make a double infinite sequence of circles on both sides of the initial circle.) In such a chain, the relation $h_n = 2nr_n$, for $n = 0, 1, 2, 3, \ldots$, of **Theorem 2.1**, is not valid and needs to be adjusted.

*Results (d) of **Theorem 2.1** and the following **Remark (c)** are valid for any complete Pappus chain.* Again, the points of tangency of the circles of the chain lie in the circumference of a circle with diameter FM on the center line FH. The centers of the circles of the chain lie on an same ellipse as before.

In both **Figures 2.6** and **2.7**, we choose a circle that intersects the center line and name it (K_0). The center K_0 may be above or below or on the center line. We let d be the singed distance of the center K_0 from the center line UV. **We consider** $d > 0$ when K_0 is above the center line UV and $d < 0$ if it is below. (If $d = 0$, then we have the classical Pappus chain.)

In this situation, we let again $r = \dfrac{R_U}{R_V}$, and we also let $s = \dfrac{d}{2R_V}$. Then, tracing the proof of **Theorem 2.1.3**, we derive the following analogous conclusions as we state them below:

(I) For the circles of the chain that are above the center line UV we have:

[(a'), analogous to (a)] Again

$$R' = R = \frac{1-r}{r} R_V, \quad \text{and} \tag{2.2}$$

$$r_n = \frac{r(1-r)}{r + [n(1-r) + sr]^2} R_V, \quad n = 0, \pm 1, \pm 2, \pm 3, \ldots . \tag{2.3}$$

We observe that the possible values of d are

$$-(1-r)R_V = -r_0 \le d \le r_0 = (1-r)R_V, \quad (n = 0, \ s = 0).$$

If $d = \pm r_0 \iff 1 - r = \pm 2s(1 + rs^2)$, then the circle (K_0, r_0) is tangent to the center line.

[(b') and (c'), analogous to (b) and (c)] For all $n = 0, \pm 1, \pm 2, \pm 3, \ldots$, we have

$$h_n = 2nr_n + \frac{2sr^2 R_V}{r + [n(1-r) + sr]^2}, \qquad FP_n = \frac{r(1+r)R_V}{r + [n(1-r) + sr]^2}.$$

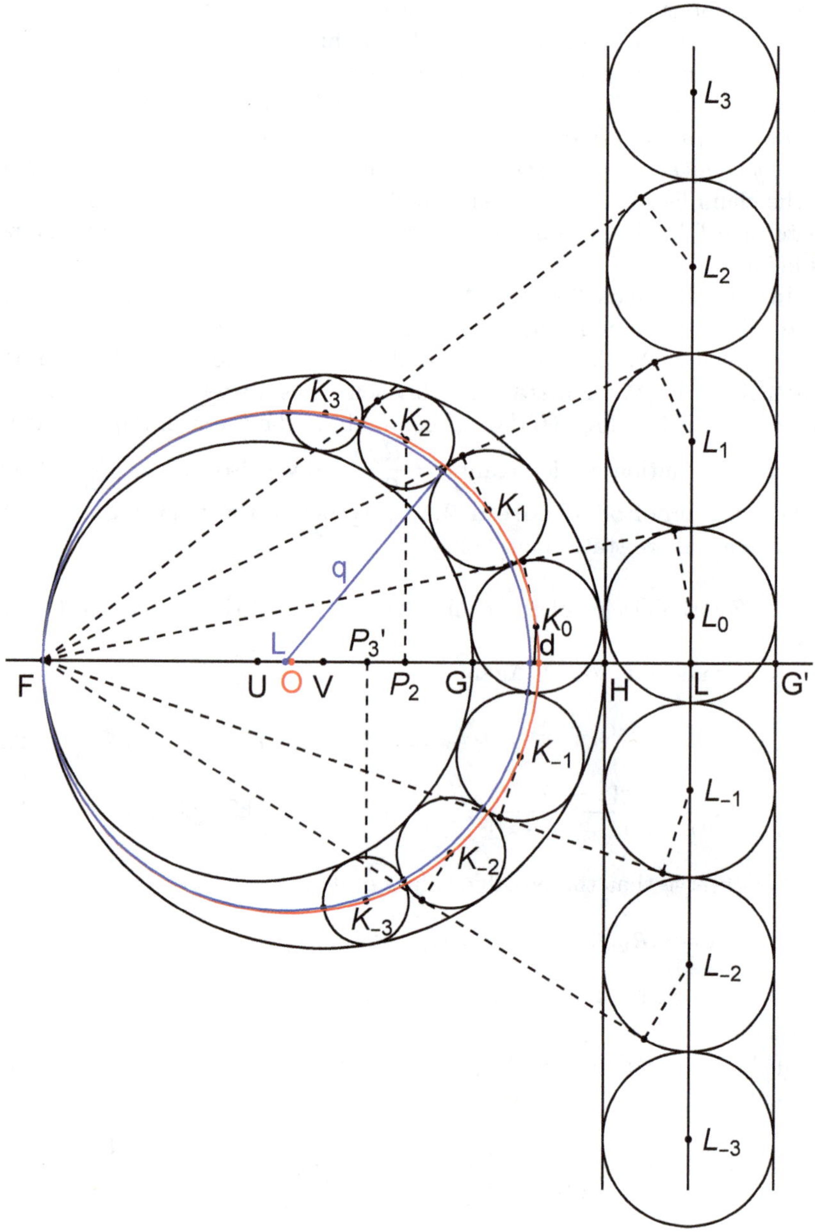

Figure 2.7: Complete General Pappus Chain

(II) For the circles of the chain which are below the center line UV the formulae are valid if instead of d we put $-d$, and so we replace s by $-s$, and then we replace h_n by $-h_n$ in the coordinates of the points of the centers $K_n = (x_n, y_n)$, for
$n = 0, \pm 1, \pm 2, \pm 3, \ldots$.

So, from these formulae, we can write down the analogous coordinates (x_n, y_n) of the centers of the circles of the chain, as we did before, if we consider the point $F = (0,0)$ to be the origin. With $r = \dfrac{R_U}{R_V}$ and $s = \dfrac{d}{2R_V}$, and for $n = 0, \pm 1, \pm 2, \pm 3, \ldots$, we find

$$x_n = R_V \frac{r(1+r)}{r + [n(1-r) + sr]^2}, \qquad y_n = 2R_V \frac{nr(1-r) + sr^2}{r + [n(1-r) + sr]^2}.$$

In fact, with $d = 0 \Longleftrightarrow s = 0$, we obtain the formulae stated before for the classical Pappus chain in **Theorem 2.1.3.**

The above formulae for r_n and (x_n, y_n) extend (generalize) Archimedes's formula 2.1 to any general classical Pappus chain.

(III) A property of these Pappus chains follows immediate from the formulae. Consider the tangent line to the base circles (U, R_U) and (V, R_V) $(R_V > R_U)$ at their common tangent point F. Then, for any circle of the chain (K_n, r_n) [or (K_{-n}, r_{-n})] the distance d_n (or d_{-n}) of K_n (or K_{-n}) from the tangent line is $d_n = FP_n$ (or $d_{-n} = FP_{-n}$). Using the formulae stated above, we see that the following ratio is constant:

$$\frac{r_n}{d_n} = \frac{1-r}{1+r} = \frac{r_{-n}}{d_{-n}} = \frac{R_V - R_U}{R_V + R_U} = \text{constant, for all } n = 0,\ 1,\ 2,\ 3,\ \ldots .$$

Remark 1: The ellipse that contains the centers of the circles of the Pappus chain is the geometrical locus of the centers of all circles that are tangent to the base circles (U, R_U) and (V, R_V), as in **Figures 2.1, 2.2, 2.3, 2.6,** and **2.7**, and not just of the centers of the circles of the chain. This ellipse, as aforementioned, has foci the center U and V of the base circles and major axis

$$M = \frac{R_U + R_V}{2},$$

and so the minor axis is

$$m = \sqrt{R_U P_V}.$$

Thus, the ellipse is completely determined.

Remark 2: The case in which the first circle of the chain $\mathcal{C}(K_0, r_0)$ is tangent to the center line FH, at a point A, is shown in **Figure 2.8**. Notice that in this configuration, we have the condition $R_U < R_V \leq 2R_U < 2R_V$.

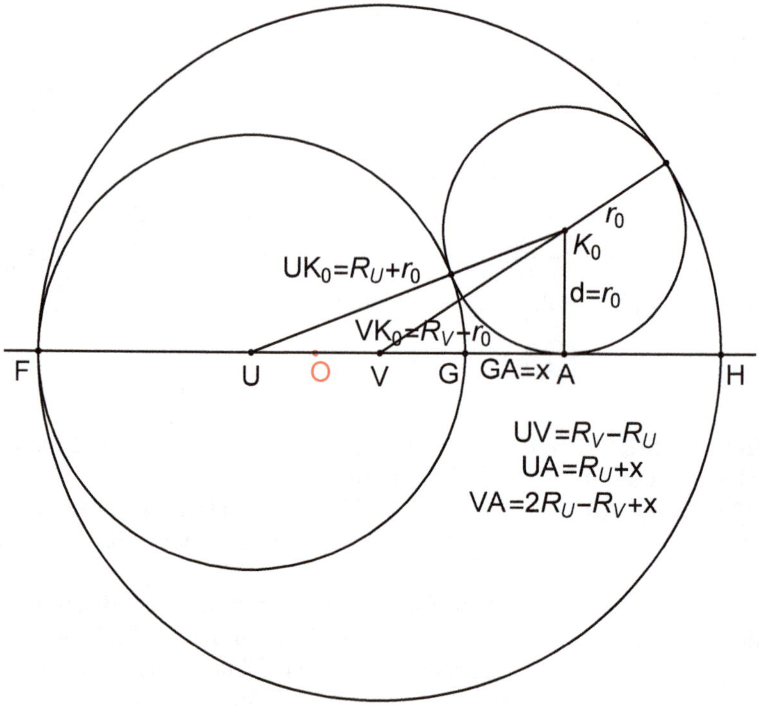

Figure 2.8: For Special Pappus Chain

Then $d = K_0A = r_0 \perp FH$. We let $x := GA$, and by the Pythagorean Theorem, we have:

$$(R_U + x)^2 + r_0^2 = (R_U + r_0)^2 \quad \text{and} \quad (2R_U - R_V + x)^2 + r_0^2 = (R_V - r_0)^2.$$

We solve and find

$$x = \frac{2R_U(R_V - R_U)}{R_U + R_V} \quad \text{and} \quad r_0 = \frac{4R_U R_V(R_V - R_U)}{(R_U + R_V)^2}.$$

Therefore, when $R_U < R_V \leq 2R_U < 2R_V$, the limits of the distance d are

$$-\frac{4R_U R_V(R_V - R_U)}{(R_U + R_V)^2} \leq d \leq \frac{4R_U R_V(R_V - R_U)}{(R_U + R_V)^2}.$$

Exercise: With $R_V > 0$ given fixed, check how the first circle of the chain $\mathcal{C}(K_0, r_0)$ behaves, in general, i.e., how x and r_0 evolve, as R_U moves in the interval $[0, R_V]$ and check the limits of the corresponding d. Distinguish and study the special cases.

2.3 Another Pappus Chain in an Arbelo

We consider the arbelo formed by the semicircles $(V,\ r_v = VF = VH)$, $(U,\ r_u = UF = UG)$, and $(W,\ r_w = WG = WH)$, similar to the arbelo of **Theorem 2.1.3**. Inside this arbelo, we now consider the Pappus chain that approaches G (and not F or H, as in this Theorem in the previous section). For the circles of the chain, we want to find a formula for their radii, and their coordinates (X_i, Y_i) with respect to point G, as origin, in terms of the given radii r_u and r_w. (Of course, $r_v = r_u + r_w$.) **Figure 2.9.**

The first circle of the chain, (K_1, r_1) is also the first circle of the chain in **Theorem 2.1.3**, in which we have proved Archimedes formula, so that, its radius and its coordinates with origin the point $F = (0, 0)$, are:

$$r_1 = \frac{(r_u + r_w)r_u r_w}{r_u^2 + r_w^2 + r_u r_w}.$$

$$x_1 = \frac{r_u(r_u + r_w)(2r_u + r_w)}{r_u^2 + r_w^2 + r_u r_w}, \qquad y_1 = \frac{2(r_u + r_w)r_u r_w}{r_u^2 + r_w^2 + r_u r_w}.$$

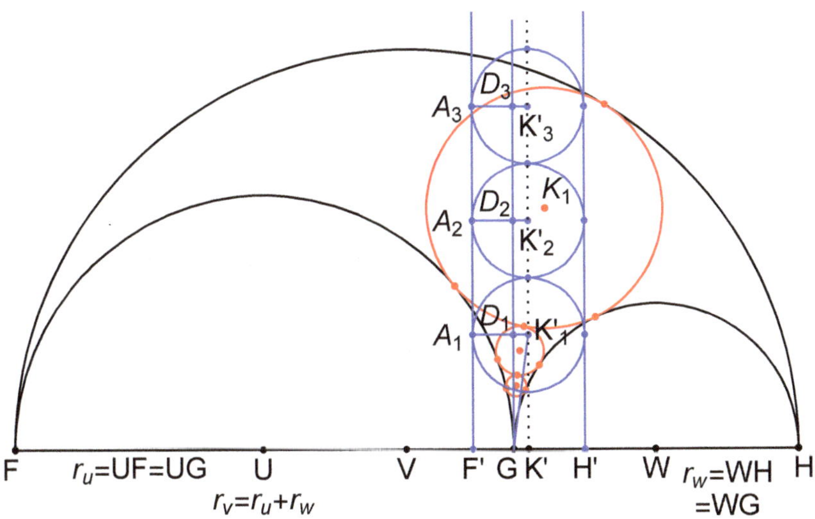

Figure 2.9: Special Pappus Chain in an Arbelo

We use inversion with power $c = r_w^2$ and center G to invert the two semicircles (U, r_u) and (W, r_w) and the circles of the chain (red circles). The two semicircles invert into the half lines perpendicular to the center line UW at the points F' and G' the inverses of the points F and G. Then

$$GF' \cdot 2r_u = r_w^2 \quad \Longrightarrow \quad GF' = \frac{r_w^2}{2r_u},$$
$$GH' \cdot 2r_w = r_w^2 \quad \Longrightarrow \quad GH' = \frac{r_w}{2},$$

The circles of the chain are inverted to the equal circles (K_i), $i = 1, 2, 3, \ldots$, tangent to each other successively and tangent to the two half lines. So, they have equal radii

$$\rho = \frac{F'H'}{2} = \frac{GF' + GH'}{2} = \frac{r_w^2}{2r_u} + \frac{r_w}{2} = \frac{r_w(r_u + r_w)}{4r_u}.$$

The x-coordinates of their centers K_i, $i = 1, 2, 3, \ldots$, with respect to G as origin, are **the same** for all circles (K_i), and given that the x-coordinate of G, (x_i), with respect to $F = (0,0)$, **they are all equal** to

$$x' = x_1 - 2r_u = \frac{r_u r_w (r_u - r_w)}{r_u^2 + r_u r_w + r_w^2}, \quad i = 1, 2, 3, \ldots.$$

For the y-coordinates we observe that

$$A_1A_2 = A_2A_3 = \ldots = 2\rho = \frac{r_w(r_u + r_w)}{2r_u},$$

the diameters of the equal circles. Next, we also prove that

$$F'A_1 = \frac{r_w(r_u + r_w)}{2r_u},$$

and so, the y-coordinates of the centers K_i, $i = 1, 2, 3, \ldots$, are

$$y_i' = \frac{i \cdot r_w(r_u + r_w)}{2r_u}, \quad i = 1, 2, 3, \ldots.$$

Since the circle (K_1, r_1) is also mapped to the circle (K_1', ρ) under the homothety with center G and homothetic ratio

$$\lambda = \frac{\rho}{r_1} = \frac{r_u^2 + r_u r_w + r_w^2}{4r_u^2},$$

we have that

$$F'A_1 = \lambda \cdot y_1 = \frac{r_w(r_u + r_w)}{2r_u} = 2\rho.$$

Then, the power of G with respect to circle (K_i'), is

$$\mathfrak{P}_i = x'^2 + y_i^2 - \rho^2 = \ldots = \frac{r_w^2[i^2(r_u + r_w)^2 - r_u r_w]}{4r_u^2}, \ i = 1,\, 2,\, 3,\, \ldots,$$

and so the homothetic ratio of the homothety with center G that maps (K_i') to the chain circle (K_i) is

$$\lambda_i = \frac{c}{\mathfrak{P}_i} = \frac{r_w^2}{\mathfrak{P}_i} = \frac{4r_u^2}{i^2 \cdot (r_u + r_w)^2 - r_u r_w}.$$

Then, the radii and if $G = (0, 0)$, the coordinates of the center of the circles of the chain (K_i), $i = 1,\, 2,\, 3,\, \ldots$, are:

$$r_i = \lambda_i \cdot \rho = \frac{r_u r_w (r_u + r_w)}{i^2 (r_u + r_w)^2 - r_u r_w},$$

$$X_i = \lambda_i \cdot x' = \frac{r_u r_w (r_u - r_w)}{i^2 (r_u + r_w)^2 - r_u r_w},$$

$$Y_i = \lambda_i \cdot y_i' = \frac{2i r_u r_w (r_u + r_w)}{i^2 (r_u + r_w)^2 - r_u r_w} = 2i r_i.$$

These formulae are the **Archimedes type formulae** for the new Pappus chain that approaches the point G (and not F).

Remarks: (1) The formula for the Y_i coordinate of (K_i), $i = 1,\, 2,\, 3,\, \ldots$, proves an analogous formula to **Pappus Ancient Theorem, 2.1.1**.

(2) With the help of the Theorem of Descartes and mathematical induction, we can prove again that for any circle (K_n, r_n), $n = 1,\, 2,\, 3,\, \ldots$, the radius is

$$r_n = \frac{(r_u + r_w) r_u r_w}{n^2 (r_u + r_w)^2 - r_u r_w}.$$

The formula is true for $n = 1$. [It can be proven independently by using Stewart's Theorem. See **Remark (d)** after **Theorem 2.1.3**.] We next assume that the formula is true for n and we will deduce it for $n + 1$. The Theorem of Descartes says that

$$2\left(\frac{1}{r_{n+1}^2} + \frac{1}{r_n^2} + \frac{1}{r_u^2} + \frac{1}{r_w^2}\right) = \left(\frac{1}{r_{n+1}} + \frac{1}{r_n} + \frac{1}{r_u} + \frac{1}{r_w}\right)^2.$$

We solve this equation to find

$$\frac{1}{r_{n+1}} = \frac{1}{r_n} + \frac{1}{r_u} + \frac{1}{r_w} \pm 2\sqrt{\frac{1}{r_u r_w} + \frac{1}{r_u r_n} + \frac{1}{r_w r_n}}$$

We must use the solution with the +, and we plug in the assumed formula for r_n. After simplifying, we obtain

$$r_{n+1} = \frac{(r_u + r_w)r_u r_w}{(n+1)^2(r_u + r_w)^2 - r_u r_w},$$

and thus the formula is proven. [This method can be used to prove the formula for r_n in **Theorem 2.1.3**, or in **Remark (d)** after it.]

(3) Since $r_u \geq r_w$, and for any n, $UK_n = WK_n = r_u - r_w \geq 0$ fixed, the centers of the circles of this chain are located on one branch of the hyperbola with foci U and W and major axis $r_u - r_w$. This hyperbola passes through G. If $r_u = r_w$, the hyperbola becomes the straight line through G and perpendicular to FH.

Another interesting case of similar chain that can be studied is the one in which we replace the large semicircle $(V, r_u + r_w)$ by the common tangent of the semicircles (U, r_u) and (W, r_w). We can make three chains, each of which approaches one of the points of tangency. In **Figure 2.10**, we indicate the chain that approaches the point G. (The reader can make and study the other two chains that approach the points T_1 and / or T_2.)

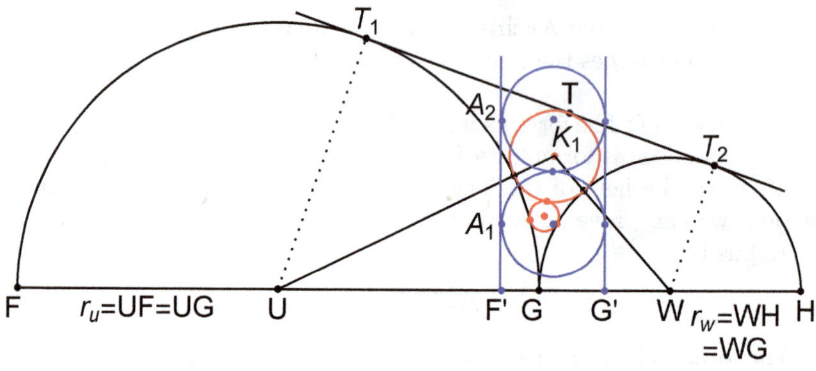

Figure 2.10: Another Special Pappus Chain

These chains can be study in the same way as the chain we have just studied above (and also study the Proof of **Descartes Theorem, 4.6.1**) as long as we find the radius and the coordinates of the center of the first circle of the chain (K_1, r_1). These elements can be figured out before of any inversion.

First, we use **Descartes Theorem, 4.6.1**, with radius of the tangent line ∞, and we find that the radius of the first circle of the chain (K_1, r_1)

is

$$r_1 = \frac{r_u r_w}{\left(\sqrt{r_u} + \sqrt{r_w}\right)^2} \left(\leq \frac{\sqrt{r_u}\sqrt{r_w}}{4}\right).$$

Hence, we know the sides of the triangle UK_1W. The semi-perimeter of this triangle is $r_u + r_w + r_1$. So, using the well known Heron's formula (see also **Chapter 4, equation 4.1**), the distance of K_1 from the line FH is the height

$$y = \frac{2}{r_u + r_w}\sqrt{(r_u + r_w + r_1) r_u r_w r_1}.$$

Then, the projection of UK_1 onto FH is equal to

$$d = \sqrt{(r_u + r_1)^2 - y^2}.$$

So, if we consider point G as the origin, the x-coordinate of K_1 is $x = d - r_u$. Hence, we determine $K_1 = (x, y)$.

Finally, we have completely determined the first circle of the chain, (K_1, r_1). But we still need to figure out the length $F'A_1$. This is done in the same way as we have done in the proof of the **Descartes Theorem, 4.6.1**. After this is found, the remaining work goes along the lines of the study of the previous chain, using inversion.

Exercise: (a) Replace the tangent line T_1T_2 in **Figure 2.10**, by a circle externally tangent to the two semicircle (or you can make them whole circles), and then construct and study the three possible chain in the non-common area enclosed by them.

(b) Use the **Descartes Theorem, 4.6.1**, to compute the radius of the first circle of the chains and then use it again to create a recursive sequence for the radii of the circles of each of the sequences.

Many chains can be studied as long as we can determine the first circle of the chain. For example, in **Figure 2.11**, we have the **Archimedean Twin Circles** in an arbelo, separated by the common tangent of the smaller semicircles at their point of tangency G. (This figure is well known and studied with its properties in the bibliography. For example, see Roussos 2022, pages 187-188, and 425-428, and other books.)

These twin circles have equal radii

$$r_1 = r_2 = \frac{r_u r_w}{r_u + r_w} \left(\leq \frac{\sqrt{r_u}\sqrt{r_w}}{2}\right).$$

(For the proof, see the above bibliography, or provide your own proof. Careful: You cannot use Descartes' Theorem to prove this known result. Why?) From the triangles WK_1V and UK_2V, we can find the coordinates of the centers in terms of r_u and r_w and origin the point G. Then we

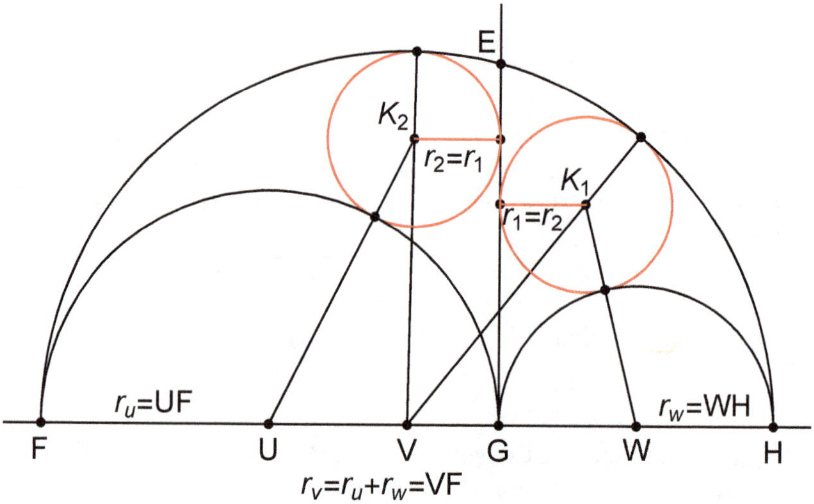

Figure 2.11: Archimedean Twin Circles in an Arbelo

continue with inversions with center G, etc. We can consider six cases of chains: Two approaching G, two E, one F, and one H. (The readers may want to work with these chains and produce results analogous to those obtained before. You can also make recursive sequences for the radii, using the Descartes Theorem and starting with the above formula for $r_1 = r_2$.)

2.4 P. C. for Externally Tangent Base Circles

There are two cases essentially: (1) Pappus chains between two externally tangent base circles. (2) Pappus chains between a circle and one of its tangent lines as base circles. (Pappus chains between two parallel straight lines are not interested and we leave them as an exercise to the readers.)

Case (1): This is depicted in **Figure 2.12**. The black circles (O_1, R_1) and (O_2, R_2) are the base circles, and the red circles are the Pappus chain. These chains are infinite.

To study this chain, we use an inversion with center the point of tangency T of the base circles and any power, or find the most convenient power $c > 0$. The images of the base circles under such an inversion are two parallel lines perpendicular to the center line. The chain circles transform to equal tangent circles between these two parallel lines. Their centers and points of tangency lie on mid-parallel line, etc. If we draw the common tangent of the two base circles at their point of tangency and we let d_n be the distance of the center K_n of any circle (K_n, r_n) of the chain from this tangent line, in this situation we easily find

$$\frac{r_n}{d_n} = \frac{R_1 + R_2}{|R_1 - R_2|}.$$

This is the reciprocal outcome of the analogous result **(III)**, examined in the previous section where the base circles were internally tangent.

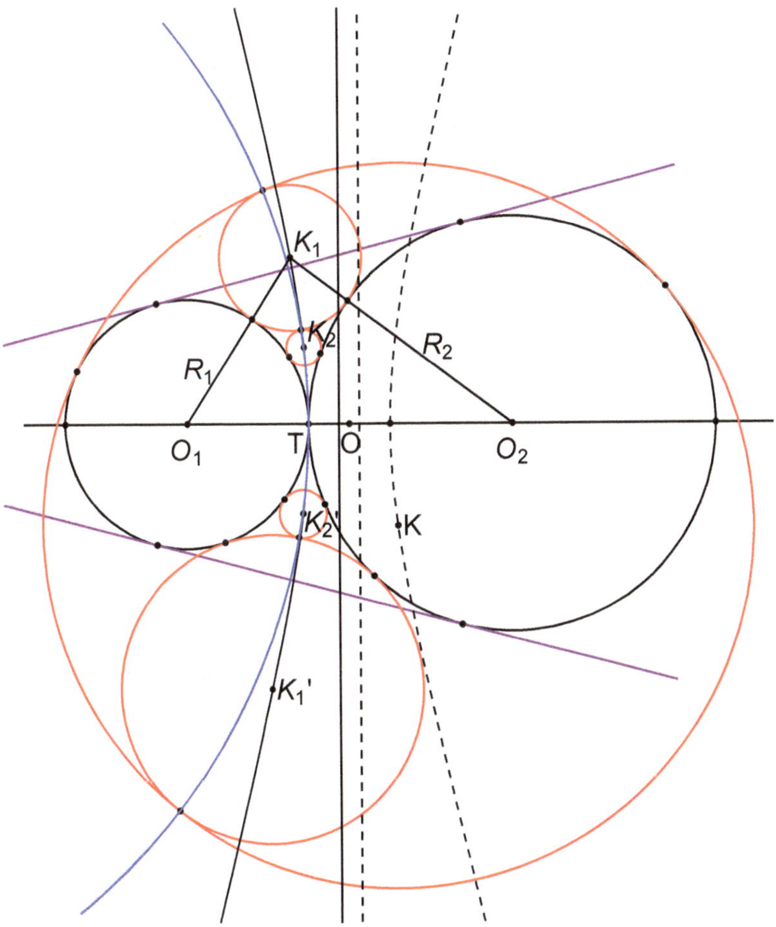

Figure 2.12: Pappus Chain for Externally Tangent Base Circles

In this case, the centers of the chain circles lie on a hyperbola with foci the centers O_1 and O_2 of the base circles. From **Figure 2.12**, it is easy to see that the difference of the distances of the center of any circle of the chain from the centers of the base circles is equal to $|R_1 - R_2|$. So, this hyperbola is completely determined. When the base circles are equal, i.e., $|R_1 - R_2| = 0$, the hyperbola degenerates to the common tangent line at the point of tangency of the two base circles, fact seen directly.

We observe:

(1) We obtain two groups of infinitely many circles of the chain as we move towards the point of tangency T from either side.

(2) If in one of these groups, one circle intersects the common tangent located above or below the common point of tangency of the two base circles, then we can either stop there, or we can find a circle that is internally tangent with this circle and the base circles. Then, we can continue the chain in the other side of T by first constructing the circle tangent to this new circle and to the base circles. This is the situation in **Figure 2.12.** But, notice that the centers of the small circles of the chain lie in the branch of the hyperbola that passes through the point of tangency T, and the center K of this new big circle lies on the other branch. (If you do not see this now, study **Section 3.8.**)

(3) If a circle of the chain is tangent to one of the common tangent lines of the two base circles, then this sequence of tangent circles stops with the tangent line and cannot be continued to the other side of the point of tangency, as we do in (2). (**Figure 2.13.**)

(4) Again the points of tangency of the circles of the chain lie on a circle (the blue arc in the figure, too big to include it all in this figure). This circle intersects the circles of the chain orthogonally, passes through the point T and has center L on the center line O_1O_2. Its radius is computed to be

$$q = \frac{2R_1R_2}{|R_1 - R_2|}.$$

Thus, this circle is completely determined. If the two base circles are equal,

$$R_1 = R_2 \quad \text{then} \quad q = \infty,$$

and the blue circle degenerates to the common tangent line of the two base circles through their point of tangency, as can be seen directly.

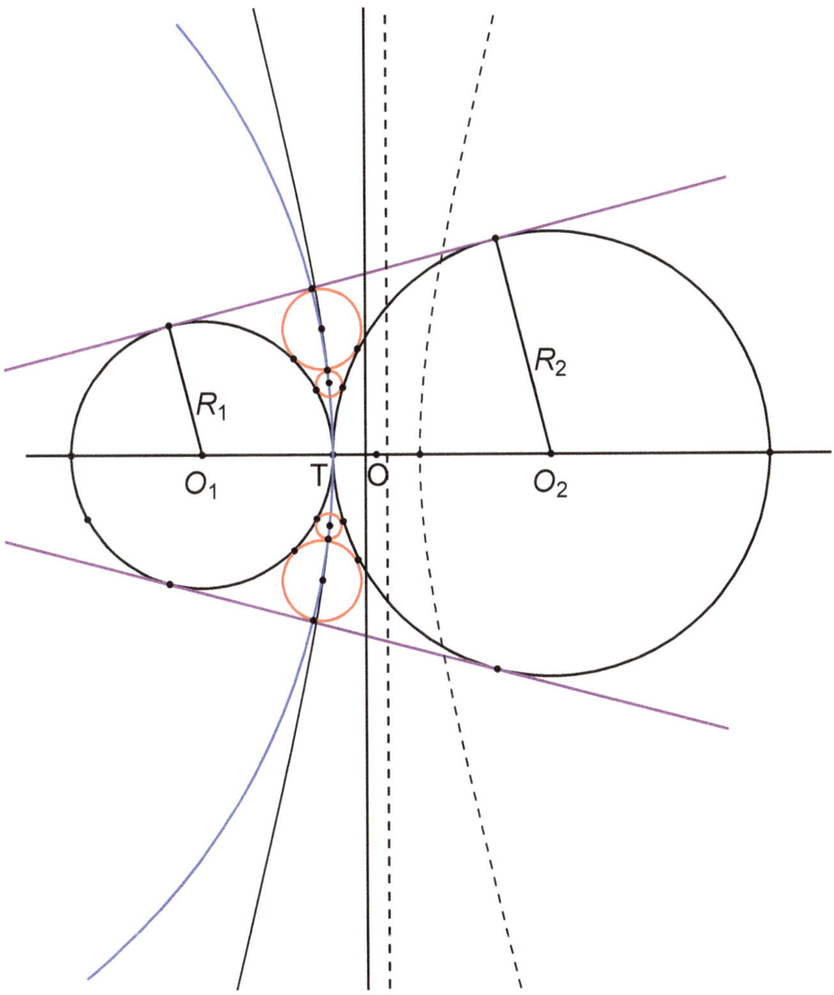

Figure 2.13: Two Pappus Chains for Externally Tangent Base Circles

(5) If the base circles are equal then the two common tangents not through their point of tangency, are parallel; viewed as circles they are tangent at their infinity point. In case that (3) above occurs, we can say that the two chains of circles are connected by means of these two parallel common tangents. (**Figure 2.14.**) In this case, we have that $R_1 - R_2 = 0$ and so the hyperbola and the blue circle degenerate to the same straight line, which is the common tangent to the equal base circles at their point of tangency. This tangent is also the midpoint perpendicular to the center segment of the base circles.

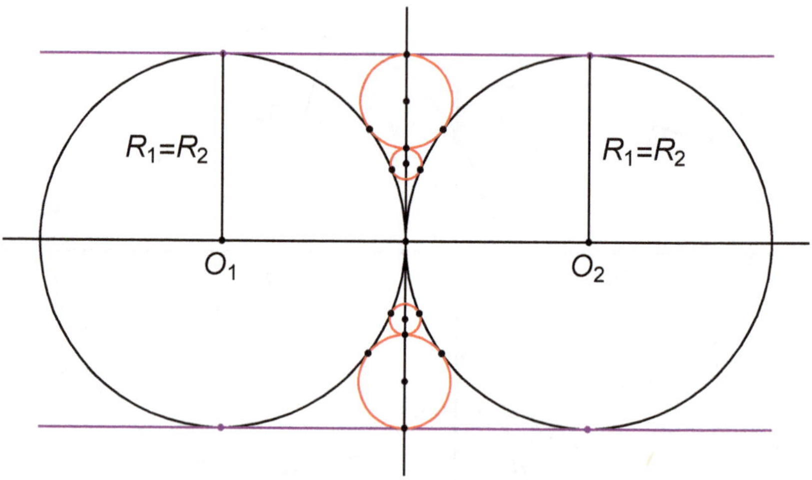

Figure 2.14: Pappus Chain for Externally Tangent Equal Base Circles. (The two external tangents are parallel and part of the chain.)

(6) To find the radius of a circle (K_n, r_n) of the chain, the data R_1, R_2 and O_1O_2 are not sufficient. This is because the first circle of the chain can be placed arbitrarily in the exterior of the circles. Of course we know that two tangent circles (O_1, r_1) and (O_2, r_1) at the point T, can be transformed to two parallel lines l_1 and l_2, by an inversion with center T and any power $c \neq 0$, **Figure 2.15**. Then, the circles (K_n, r_n) of the chain are mapped to equal circles tangent to these parallel lines and to each other successively. To continue in this way to figure out the radii of the circles of the chain is complicated and cumbersome. Someone may want to finish it, under the condition that he knows everything about the position and radius of the circle (K_1, r_1) of the chain. In such a chain, a common tangent of the two

base circles may be included, and or a tangent circle that contains the two base circles in its interior.

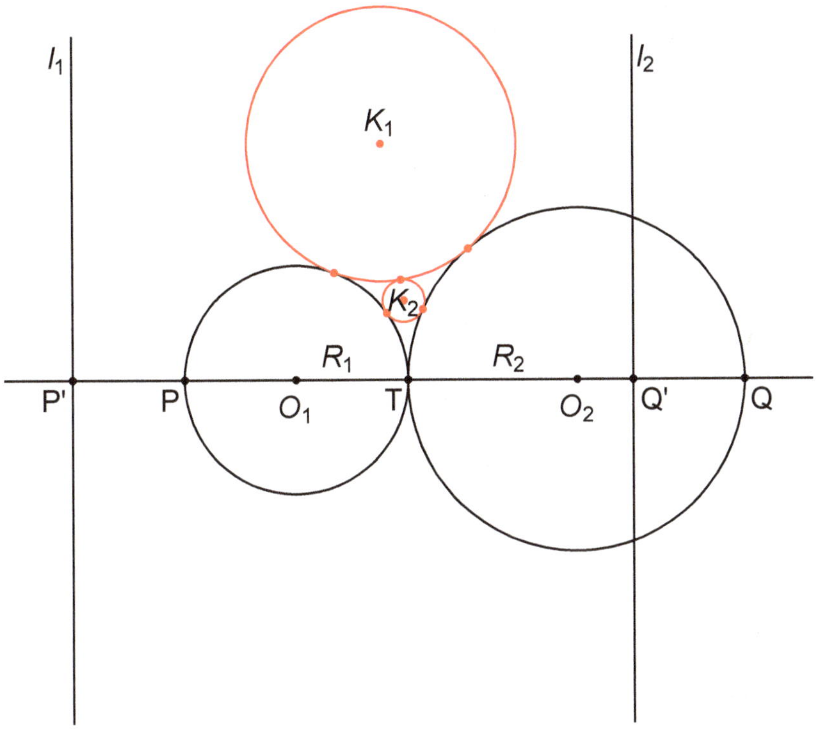

Figure 2.15: Tangent Circles Inverted to Parallel Lines

We can use other methods to find the radii of the circles of the chain. (a) If we know the projection H_n of the center K_n of the circle on the line O_1O_2, we can find the radius r_n, as follows. The midpoint M of the segment O_1O_2 is known, and so the segment H_nM is known. Now, we use the theorem of medians in a triangle to get:

$$(R_1 + r_n)^2 - (R_2 + r_n)^2 = 2(O_1O_2)\overline{H_nM} = 2(R_1 + R_2)\overline{H_nM}$$

$$\Longrightarrow \quad r_n = \frac{(R_1 + R_2)(2\overline{H_nM} - R_1 + R_2)}{2(R_1 - R_2)}, \quad \text{if} \quad R_1 \neq R_2.$$

If $R_1 = R_2$, then $H_nM = 0$ and the formula takes on the undeterminate form $\frac{0}{0}$. In this case the hyperbola degenerates to the common tangent line of the two base circles at the common point of tangency. (**Figure 2.14.**)

(b) If we know the radius of one circle of the chain, then a better way to find the radii of the circles of the chain, one by one, is to use the **Theorem of Descartes**, successively. (This Theorem states the necessary and sufficient condition that in four given circles any one of them is tangent to the other three. See bibliography, for example: Coxeter, H. S. M., *Introduction to Geometry*, pp. 13-15. Pedoe D., *Geometry a Comprehensive Course*, pp. 157-158. See also the hint given at the end of **Section 3.8**, **Section 4.6** for other proofs, and **Theorem 4.6.1**, **Chapter 4**, proved by inversion.)

Let us suppose that we know the radius r_1 of one circle of the chain. Then, we apply **equation 4.7**, or **4.8**, the next radius r_2 is found by solving **Descartes' formula** for r_2:

$$2\left(\frac{1}{R_1^2}+\frac{1}{R_2^2}+\frac{1}{r_1^2}+\frac{1}{r_2^2}\right)=\left(\frac{1}{R_1}+\frac{1}{R_2}+\frac{1}{r_1}+\frac{1}{r_2}\right)^2 \iff$$
$$\frac{1}{r_2}=\left(\frac{1}{R_1}+\frac{1}{R_2}+\frac{1}{r_1}\right)\pm 2\sqrt{\frac{1}{R_1R_2}+\frac{1}{R_2r_1}+\frac{1}{R_1r_1}}.$$

If the solutions with the $(+)$ is always positive and gives the radius r_2 in **Figure 2.12**. If the solution with the $(-)$ is zero, then this corresponds to one common tangent of the base circles, not through the common point of tangency. If this solution is positive, then this will give the radius of the circle of the chain on the other side of the circle (K_1, r_1) that has the same type of tangency (external tangency in this figure).

But, if the solution with the $(-)$ is negative, then its absolute value is the radius of the big red circle, as shown in this figure. This is seen by the fact that this circle has internal tangency with both the base circles and the circle (K_1, r_1), and so, its radius should be computed by the following adjusted formula (as Descartes' Theorem also provides):

$$2\left(\frac{1}{R_1^2}+\frac{1}{R_2^2}+\frac{1}{r_1^2}+\frac{1}{r^2}\right)=\left(\frac{1}{R_1}+\frac{1}{R_2}+\frac{1}{r_1}-\frac{1}{r}\right)^2 \iff$$
$$\frac{1}{r}=-\left(\frac{1}{R_1}+\frac{1}{R_2}+\frac{1}{r_1}\right)\pm 2\sqrt{\frac{1}{R_1R_2}+\frac{1}{R_2r_1}+\frac{1}{R_1r_1}}.$$

Now, the solution with the $(-)$ is always negative, and has no geometrical meaning. We accept the solution corresponding with the $(+)$, which is the absolute value of the previous expression with the $(-)$. (Study the details of Descartes' Theorem and its variations in **Chapter 4** and the bibliography.)

Next, for the other circles of the chain, we use the this method successively. The method also works when $R_1 = R_2$.

Remark: To understand the role of the hyperbola found in **Figure 2.12**, study **Section 3.8**.

Case (2): This is depicted in **Figure 2.16**. Here, we have a circle (O, R) and a tangent line to it at a point A. The red circles are the circles of the Pappus chain. These chains are again infinite.

To study such a chain, we use an inversion with center the point of tangency A of the base circle and its tangent line. A convenient power is

$$c = (2R)^2.$$

Then, the image of tangent line is itself and the image of the base circle under such an inversion is a line parallel to the tangent line at distance $2R$. The chain circles are transformed to equal tangent circles between these two parallel lines, etc. Again, using this inversion, as in **Case 1**, we prove that the points of tangency lie on a circle (the blue circle) with radius

$$q = \frac{2}{\frac{1}{R} + \frac{1}{\infty}} = \frac{2}{\frac{1}{R} + 0} = 2R,$$

and so, its center L is the diametrical point to A on the circle (O, R).

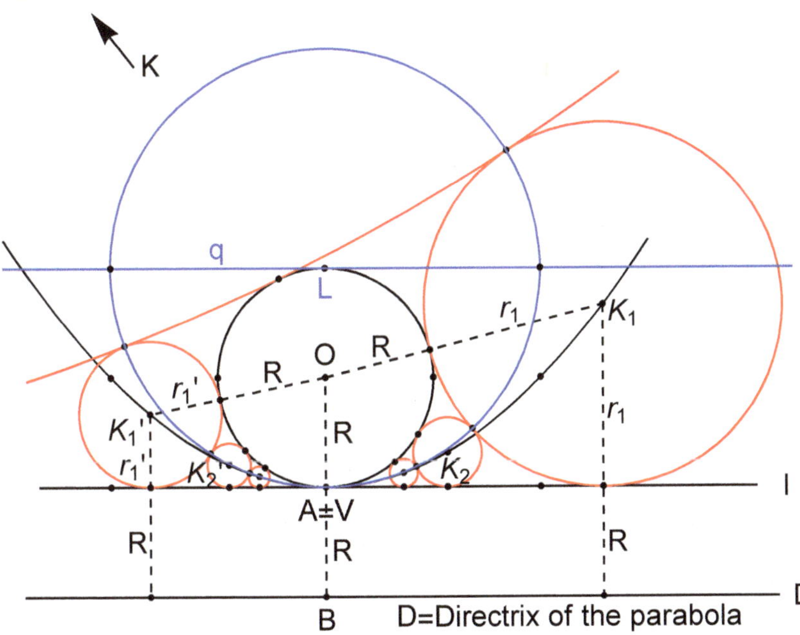

Figure 2.16: Pappus Chain for Circle and Tangent Line as Base Circles

[Another way to argue is by basic geometry, since the circle (L, q) has to pass through the point of tangency $A = V$. So $q = 2R$. Also, this circle intersects the circles of the chain (red circles) orthogonally. (Provide the details.)]

In this case, the centers of the circles of the chain lie on a parabola with focus the center O of the base circle, vertex the point of tangency A, and directrix the line parallel to the tangent line at distance R and on the half-plane that O does not lie. Looking at the figure, we see that the distance of the center of any circle of the chain from the center of the base circle is equal to its distance from the indicated directrix. So, the centers are located on this parabola, which is completely determined. (To this result, study also **Section 3.8**.)

We observe:

(1) We obtain two sequences of infinitely many circles of the chain as we move towards the point of tangency A from either side.

(2) If in one of these groups one circle intersects the tangent of the base circle at the point L, the diametrical point of A, then we can stop there, or we can find another circle that is tangent to this circle, and the base circle, and the tangent line of the base circle at A. Then, we can continue the chain in the other side of A by first constructing the circle tangent to this new circle, to the base circle and the given tangent line. This is the situation of in **Figure 2.16**. (This circle was to big to put it all in this figure. We have made an arc of it and indicated the direction in which its center K is located.)

(3) If a circle of the chain is tangent to the tangent line of the base circle at L, the diametrical point of A, then we view this tangent as an infinite circle and we continue the chain on the other side of A and the new group is symmetrical to the first one in the line AL. (**Figure 2.17**.)

(4) If we have two different parallel lines (instead of one circle and one of its tangent lines), then we can make an infinite chain of equal circles between them. In this case the parabola and the blue circle degenerate to the same straight line, which is the mid-parallel line to the two given parallel lines. This chain is simple and not interesting. (To see this, draw an easy figure.)

(5) We can find the radii of the circles of the chain, if we know the radius of one initial circle of the chain and then we use the **Theorem**

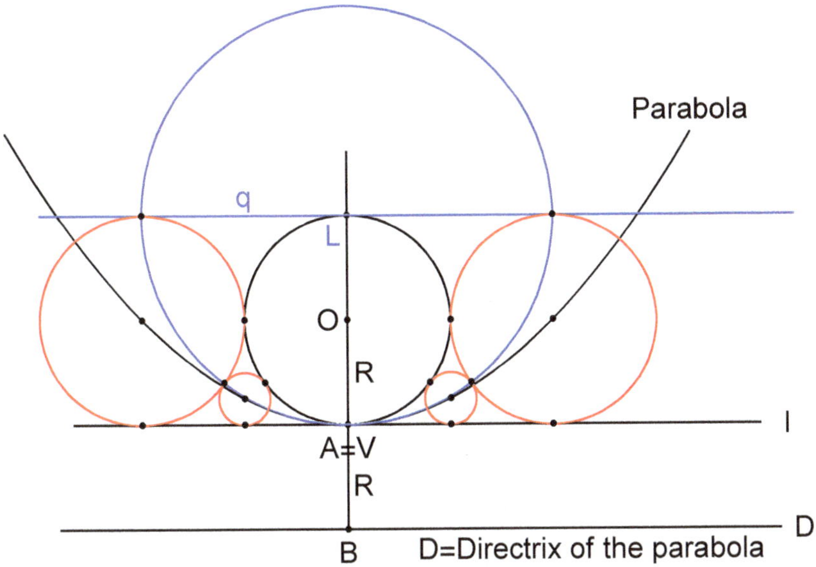

Figure 2.17: Pappus Chain for Circle and Tangent Line as Base Circles, in which the Tangent Line Parallel to the Given Tangent Line is included

of Descartes, where in this situation, we consider the straight line as a circumference of infinite radius ($R_2 = \infty$). For example, let r_1 be known and let $R_1 = R$ and $R_2 = \infty$. Then,

$$\frac{1}{R_2} = \frac{1}{\infty} = 0$$

and r_2 is found by solving the corresponding Descartes' formula for r_2:

$$2\left(\frac{1}{R^2} + 0 + \frac{1}{r_1^2} + \frac{1}{r_2^2}\right) = \left(\frac{1}{R} + 0 + \frac{1}{r_1} + \frac{1}{r_2}\right)^2 \iff$$

$$\frac{1}{r_2} = \left(\frac{1}{R} + \frac{1}{r_1}\right) \pm 2\sqrt{\frac{1}{Rr_1}}.$$

Here, both solutions are positive or one positive and the other zero.The solution with the plus (+) gives the radius r_2 in **Figure 2.16** and the solution with the minus (−) gives the radius of the circle on the other side of the circle (K_1, r_1). For the other circles, we use the same method successively. The zero ($r_2 = 0$) solution is present if and only if $R = r_1$, and then $r_2 = \infty$. This corresponds to the straight line tangent to the given circle at the point L (diametrical to A) and parallel to the given tangent line l, as seen in **Figure 2.17**.

Remarks: (1) Again, as in **Case (1)**, the circle (L, q) intersects the circles of the chain orthogonally. Therefore, all common tangents of all pairs of successive circles of a Pappus chain at the common points of tangency are radial lines of the circle (L, q), and so, concurrent at the center L.

(2) To understand the role of the parabola found in **Figures 2.16** and **2.17**, study **Section 3.8**.

(3) Any inversion with center the point of tangency $A = V$ and any power $c \neq 0$, maps the tangent line onto itself and the circle (O, R) to a straight line parallel onto the tangent line. All the circles of the chain are mapped to equal circles between these two parallel lines.

2.5 P. C. for Intersecting Base Circles

There are two cases essentially.
(1) Pappus chains between two circles that intersect at two points.
(2) Pappus chains between a circle and a straight line that intersect at two points.
(Pappus chains between two intersecting straight lines are not interested and we leave them as a simple exercise to the readers.)

Case(1): We consider two circles (O_1, R_1) and (O_2, R_2) that intersect at two points A and B. See **Figure 2.18**. Then, with these two circles as base circles (the black circles in the figure), we can construct four chains of circles, one in each of the areas:
(1) the intersection of the circles,
(2) and (3) the two disjoint parts of the circles, and
(4) the exterior of the circles. The chains are infinite, as we can construct infinite circles approaching the points A and B.
An inversion with center one of the common points, A, let us say, transforms the two base circles into two straight lines that intersect at the inverse image of the point B, and the chain circles are transformed into circles tangent to these straight lines and lying inside the four parts that these lines partition the plane. See **Figure 2.19**.

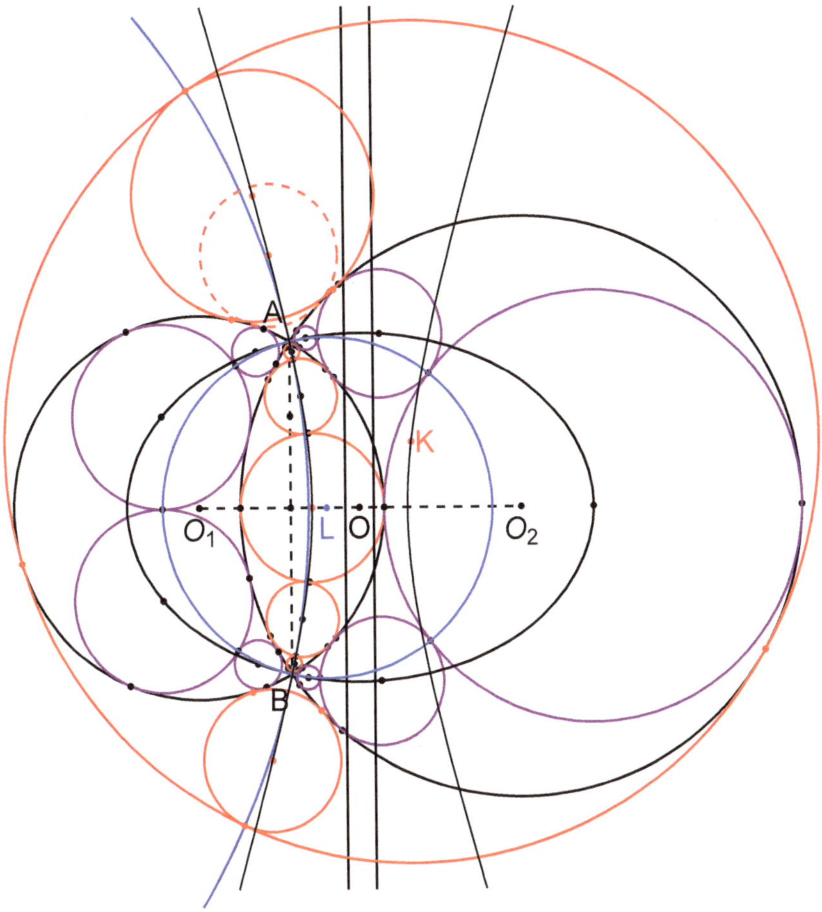

Figure 2.18: Pappus Chains for two Intersecting Base Circles

We observe the following:

(1) The chains inside the intersection and in the exterior of the base circles can be viewed as continuations of each other. The centers of these chain circles are located on a hyperbola with foci O_1 and O_2. We readily see that the difference of the distances from the points of the hyperbola to the foci O_1 and O_2 is $|R_1 - R_2|$. Thus, the hyperbola is completely determined. One branch passes through the common points A and B and the other branch contains the center K of the big circle of the exterior chain. Such a circle is possible if it can be tangent to the base circles and a circle of the chains that approach A and B.

(2) The chains inside the two disjoint parts of the base circles can also be viewed as continuation of each other. The centers of these chain circles are located on an ellipse with foci O_1 and O_1. We readily see that the sum of the distances from the points of the ellipse to the foci O_1 and O_1 is R_1+R_2. This ellipse passes through A and B and it is completely determined.

(3) In either pair of the chains, the points of tangency lie in the blue circles that intersect the circles of the respected chains orthogonally, pass through A and B and have centers on the center line O_1O_2. We them in **Figure 2.19** and ill elaborate in what follows, next.

In **Figure 2.19**, we have use an inversion of some positive power with center the point A. We have drawn the inverse of several circle of the chains and of one common tangent line of the base circles in purple color. The inverse image of this tangent is the big purple circle tangent to the two straight lines, which are the images of the base circles. (All these figures are quite involved, and so, we use several colors to make it easier to see all the properties of the chains.)

In this figure we have:

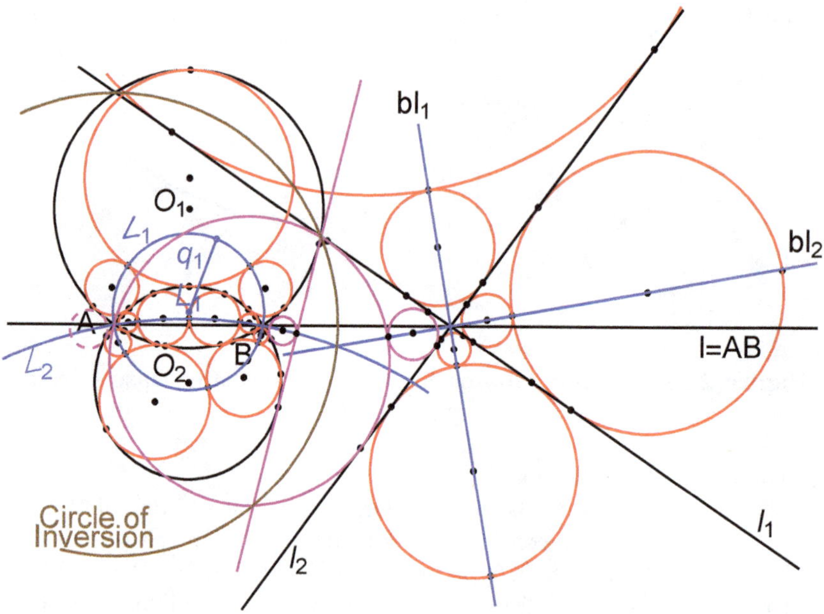

Figure 2.19: Inverse of a Pappus Chain for Intersecting Base Circles

Image of (O_1, R_1) is the line l_1.
Image of (O_2, R_2) is the line l_2.
Image of blue circle $\mathcal{L}_1$ is the line bl_1.
Image of blue circle $\mathcal{L}_2$ is the line bl_2.
Image of the point B is the common point of the above four lines.

Images of the red circles of the chains are the red circles between the lines l_1 and l_2, tangent to them, and with centers on the lines bl_1 and bl_2.

The lines bl_1, bl_2 are the bisectors of the angles of the lines l_1 and l_1.

The points of tangency and the centers of circles which are the images of the circles of the chains are on the bisectors of the straight lines l_1 and l_2 (the images of the base circles). (But remember that centers of circles are not mapped to the centers of the image circles under inversion.) The two bisectors are perpendicular. Therefore, the two blue circles (in **figures 2.18** and **2.19**) that contain the points of tangency of the circles of the chains are orthogonal and bisect the angles of the base circles.

Knowing R_1, R_2 and O_1O_2, we can find the radii of the two blue circles in the following steps.

First, we let

$$s = \frac{R_1 + R_2 + O_1O_2}{2}.$$

AB is double the height of triangle O_1AO_2 and then, by Heron's[4] formula we get that

$$AB = 2 \cdot \frac{2}{O_1O_2} \cdot \sqrt{s(s - R_1)(s - R_2)(s - O_1O_2)}.$$

Second, we compute the angle $\hat{\theta} = \widehat{O_1AO_2} = \widehat{A}$ of the triangle O_1AO_2

$$\theta := \arccos\left[\frac{R_1^2 + R_2^2 - (O_1O_2)^2}{2R_1R_2}\right].$$

Third, we find the angle $\widehat{\phi_1} := \widehat{O_1}$ of the triangle O_1AO_2, which is also one of the angles between the line AB and the tangent of the circle (O_1, R_1) at the point A (or B, as these angles are equal. See **Figure 2.20**.) We have that

$$\phi_1 := \arccos\left[\frac{R_1^2 + (O_1O_2)^2 - R_2^2}{2R_1O_1O_2}\right].$$

Similarly, the angle $\widehat{\phi_2} := \widehat{O_2}$ of the triangle O_1AO_2 is

$$\phi_2 := \arccos\left[\frac{R_2^2 + (O_1O_2)^2 - R_1^2}{2R_2O_1O_2}\right].$$

[4]Heron of Alexandria, Greek mathematician and engineer, ?-70 CE

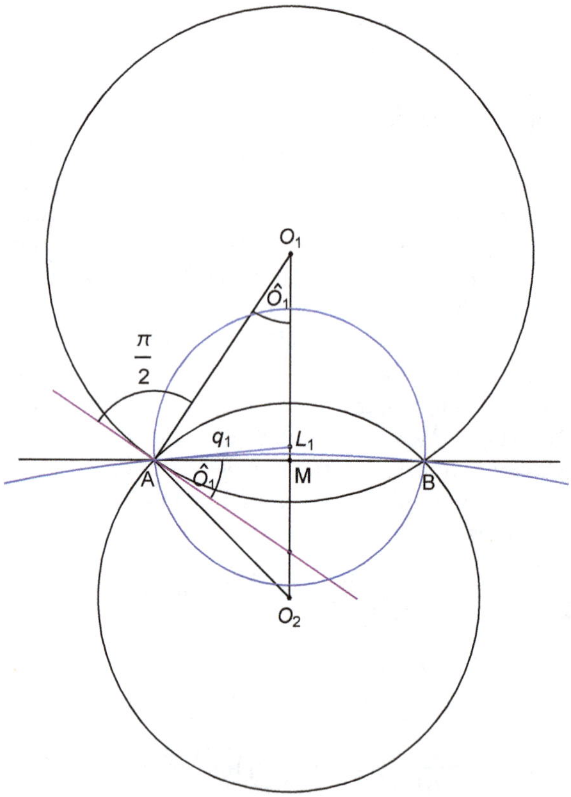

Figure 2.20: For the Determination of the two blue Circles that contain the Points of Tangency of the Chains

Then, using the properties of inversion, or easier, by the fact that the blue circles pass through the points A and B (the line AB is the common cord with the base circles and the blue circles), and bisect the angles of the base circles, we geometrically prove (check the right triangle ML_1A, in **Figure 2.20**) that the **radius** of the blue circle that contains the points of tangency of the circles of the chains inside the non-common parts is:

$$q_1 = \frac{AB/2}{\sin\left(\frac{\theta}{2} + \phi_1\right)} = \frac{AB}{2\sin\left(\frac{\theta}{2} + \phi_1\right)}.$$

Now, if M is the **midpoint** of the segment AB, then on the line O_1O_2 and in the correct side of AB, we find the **center** L_1 of the above circle, by the relation

$$ML_1 = \sqrt{q_1^2 - \frac{AB^2}{4}}.$$

Similarly, for the blue circle that contains the points of tangency of the circles of the chains inside the common and the exterior parts of the base circles, we have that the **radius** is

$$q_2 = \frac{AB}{2|\cos\left(\frac{\theta}{2} + \phi_2\right)|},$$

not drawn in **Figure 2.19**. (Too big.)

The the **center** L_2 is located on the line O_1O_2 by the relation

$$ML_2 = \sqrt{q_2^2 - \frac{AB^2}{4}}$$

on the half plane of the line AB that L_1 does not lie.

Now that we know the radii and the fact that the blue circles pass through the points A and B (and so their centers are on the center line O_1O_2, the midpoint perpendicular of AB), we can locate the centers L_1 and L_2 of the two blue circles $\mathcal{L}_1 := (L_1, q_1)$ and $\mathcal{L}_2 := (L_2, q_2)$, by a simply geometrical construction.

[Notice that if the line $l = AB$ is tangent, i.e., the base circles are externally tangent, then in this limiting position the formula for q_1 has the form $\frac{0}{0}$ and $q_2 = 0$. The radius q_1 was computed earlier in **section 2.4, Case (1).**]

Remark: The circles $\mathcal{L}_1 = (L_1, q_1)$ and $\mathcal{L}_2 = (L_2, q_2)$ intersect the circles of their associated chain orthogonally. Therefore, all common tangents of all pairs of successive circles, in each of these Pappus chains, at the common points of tangency are radial lines of the corresponding circle, and so, concurrent at its center.

Next, we study the ratio of the radius of a chain circle and the distance of its center from the line AB. In **Figure 2.21**, we consider the chains formed in the interior of two base circles (O_1, R_1) and (O_2, R_2). Again the two base circles (O_1, R_1) and (O_2, R_2) intersect at two points A and B.

For the chain in the intersection of the two base circles, we pick any circle (K, r) and we let d be the distance of K from the line AB. We let α_1 be the non-oriented angle between the circle (O_1, R_1) and the line AB and α_2 the non-oriented angle between the circles (O_2, R_2) and the line AB. [These angles are the angles between the line AB and the tangents to the base circles (O_1, R_1) and (O_2, R_2) at A, or B. So, $\alpha_1 + \alpha_2$ is one of the non-oriented angles between the two base circles.]

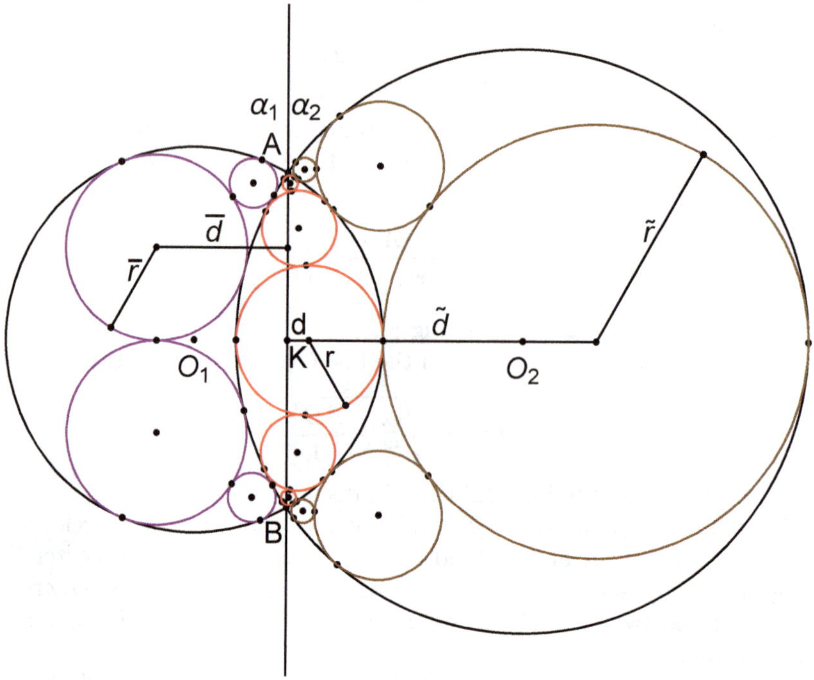

Figure 2.21: Pappus Chains in the Interior of Circles that Intersect

Inverting this chain by the inversion mention before, that is, the inversion with center A and any power $c > 0$, we partially find **Figure 2.22**, in which we have the following:

The image of B is B'.

The image of circle (O_1, R_1) is the line l'_1.

The image of circle (O_2, R_2) is the line l'_2.

The image of Line AB ia $B'A'$.

The image of circle (K, r) is the circle (L', r').

[Remember that K does not correspond to L' under the inversion. They are not inverse of each other, but homothetic by a homothety determined by the inversion, with center the same as the center of the inversion (which is A), and its homothetic ratio is

$$\lambda = \frac{|c|}{|p|},$$

where $p = AK^2 - r^2$, the power of A with respect to the circle (K, r).]

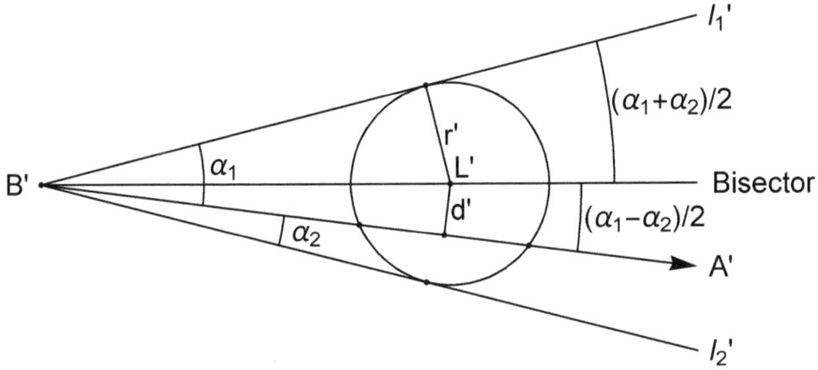

Figure 2.22: For the Pappus Chains in the Interior of Base Circles that Intersect, in Figure 2.21

We finally let d' be the distance of L' to the line $B'A'$.

Any two inverted circles inside the angle $l_1'\hat{B}'l_2'$ are homothetic with center of homothety B'. (Their centers are on the bisector of the angle.) Also, any original circle (K, r) is homothetic with the circle (L', r') with center of homothety A. Then, by the properties of homotheties, we get

$$\frac{r}{d} = \frac{r'}{d'}.$$

Also, inversions preserve the absolute values of angles.Then, using the similar right triangles inside the angle $l_1'\hat{B}'l_2'$, we obtain

$$\frac{r}{d} = \frac{r'}{d'} = \frac{\sin\left(\frac{\alpha_1+\alpha_2}{2}\right)}{\sin\left(\frac{|\alpha_1-\alpha_2|}{2}\right)},$$

the same constant for all circles of the chain.

(Using trigonometric and geometric formulae, we can express the sines in this formula in terms of R_1, R_2 and O_1O_2. Try it out as an exercise. See also Yaglom 2009.)

The **same result** is obtained **for the chain** which is located in the **exterior** of the circles, even for the big circle of the chain (if such a circle is possible) that is tangent to the extreme circles of the two chains and the base circles. (Work it out. See **Figure 2.18**. These two chains could be considered as continuation of each other.)

We can do analogous figures and work for the other two chains that are inside the disjoint parts of the base circles, as they are indicated in **Figure 2.21**. Make a figure analogous to **Figure 2.22** to represent this case and prove that

$$\frac{\bar{r}}{\bar{d}} = \frac{\bar{r}'}{\bar{d}'} = \frac{\sin\left(\frac{\pi+\alpha_1+\alpha_2}{2}\right)}{\sin\left(\frac{\pi+|\alpha_1-\alpha_2|}{2}\right)} = \frac{\cos\left(\frac{\alpha_1+\alpha_2}{2}\right)}{\cos\left(\frac{\alpha_1-\alpha_2}{2}\right)} = \frac{\tilde{r}'}{\tilde{d}'} = \frac{\tilde{r}}{\tilde{d}},$$

which is a constant, the same for all circles of both chains (one continuation of the other).

Case (2): We consider a circle (O, R) and a straight line l that intersect at two points A and B. Then, with this circle and the straight line (in black color in the figure) as base circles, we can construct four chains of circles, one in each of the four connected areas they partition the plane. The chains are infinite, as we can construct circles approaching the points A and B. See **Figures 2.23** and **2.24**, in which of which we have drawn two chains. The two chains in each pair can be considered and continuation of each other. We are going to examine some properties of these chains.

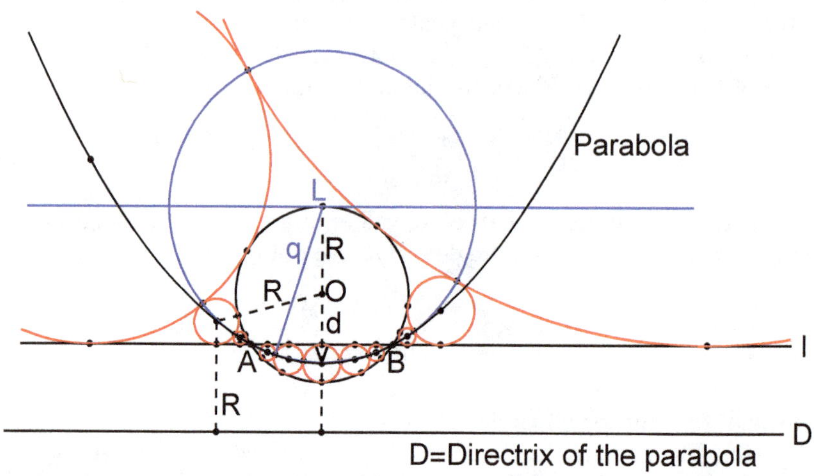

Figure 2.23: Pappus Chain for Circle and Line that Intersect

We observe the following:

(1) The pair chains in the interior and the exterior of the circle (as they are shown in the two figures), can be viewed as continuation of each other. The centers of the circles of each of these pairs of chains are located on two

parabolas with the same focus O, and directrices the straight lines parallel to the line l at distance R, one in each half plane of the line l. (From the figures, we directly see that the distance from the points of the parabola to the focus O and the respected directrix are equal.) Thus, these parabolas are completely determined. Also, the parabolas pass through the common points A and B.

(2) In the two pairs of chains, the points of tangency lie in the blue circles that intersect the circles of the respected chains orthogonally, pass through A and B and have centers on the line through O and perpendicular to l.

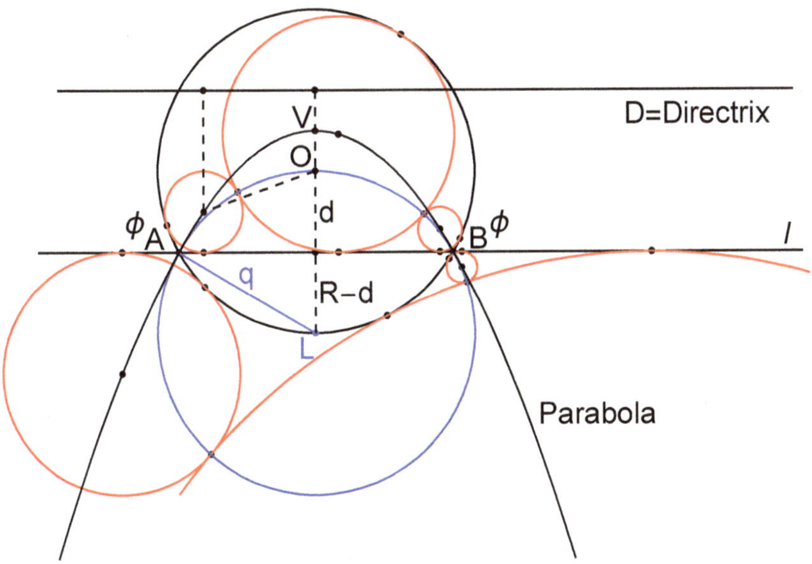

Figure 2.24: Pappus Chain for Circle and Line that Intersect

We let

$$d = \sqrt{R^2 - \frac{AB^2}{4}}$$

the distance of the center O from l and $0 \leq \phi \leq \frac{\pi}{2}$ the angle between the line l and the tangent to the circle at the point B (or A, as these angles at either A or B are equal). Using the fact that the blue circles pass through A and B we can directly prove (or by using inversion), that in the situation of **Figure 2.23**, the center of the blue circle is the point of the given base circle (O, R) on the diameter perpendicular to l and on the side that O lies.

Also this circle bisects one of the angle of the given base circle and straight line l. We observe that the radius of this circle is

$$q = \frac{R+d}{\cos\left(\frac{\phi}{2}\right)}.$$

Similarly, in the situation of **Figure 2.24**, the center of the blue circle is the point of the given circle (O, R) on the diameter perpendicular to l and on the side of L that O does not lies. We observe that the radius of this circle is

$$q = \frac{R-d}{\cos\left(\frac{\phi}{2}\right)}.$$

So, in either situation, the blue circle is completely determined since we know two points of it, A and B, and its radius. (We can compute its center L in the same we as we have done before. But, as have seen above, we have a faster way, in this situation.) Again, in either case, the circle (L, q) intersects the circles of the chain orthogonally. Therefore, all common tangents of all pairs of successive circles in each of these Pappus chains at the common points of tangency are radial lines of this circle and so concurrent at its center L.

It is possible the tangent line to the base circle (O, R) at the point L to be an infinite circle in the Pappus Chain of this present case. See **Figures 2.25** and **2.26**.

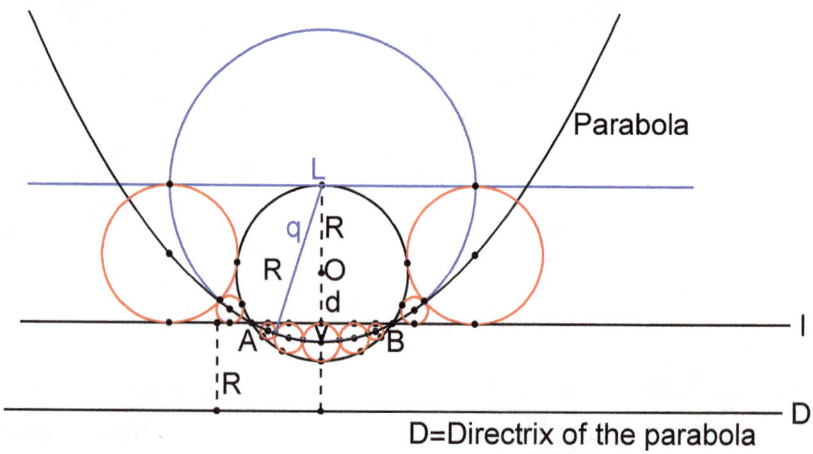

Figure 2.25: The tangent to the Base Circle at L is in the Pappus Chain

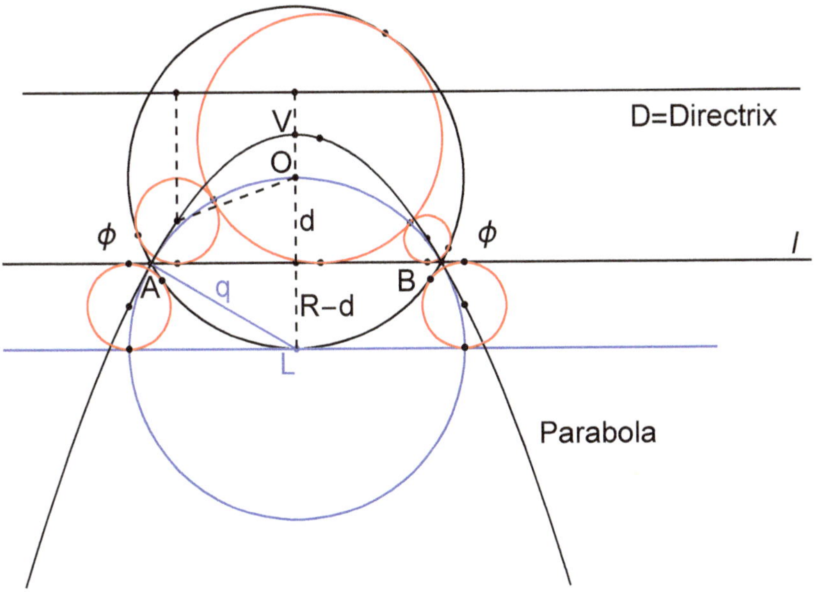

Figure 2.26: The tangent to the Base Circle at L is in the Pappus Chain

In this situation the external parts of the chain must be symmetrical in the line OL. (The internal chain does not have to be symmetrical. It can be either way.)

Remark: In this chapter we have seen that, in any Pappus chain, whereas the points of tangency of the circles of the chain are always on a specific circle or straight line, the centers of the circles of the chain are on a **conic section**:

(1) Ellipse,
(2) Parabola,
(3) Hyperbola,
(4) Straight line.

(1) occurs if the base circles are internally tangent, or when the two base circles intersect at two points.

(2) occurs if one base circle is a straight line and the other circle is either tangent to it or intersects it at two points.

(3) occurs when the base circles are externally tangent, or the two base circles intersect at two points.

(4) occurs, if the base circles are equal. Then for one of the chains the conic section degenerates to a straight line. It also occurs, if we have a chain between two parallel lines. Then the centers are on the mid-parallel straight line.

(For further extension and understanding of the connection of the conic sections with the chains of circles, study **Section 3.8**.)

Chapter 3

Steiner Chains

A chain of circles is called a **Steiner chain** (S. C., for short), if its two base circles, and or a circle with a straight line, have no common point. These chains are either finite (periodically infinite) or infinite. These chains exhibit many interesting properties that we will examine in the sequel. The finite Steiner chains are more interesting and applicable than the infinite ones.

We will examine the three interesting cases:
(1) One of the two base circles is inside the other (nested base circles).
(2) One of the base circles is a straight line.
(3) The two base circles are outside of each other.
(Steiner chains between two parallel lines are simple and not interesting, and so, we will not spend any time on them. Make a figure of such a chain and you will see their few properties readily. We have also seen that these chains may also be considered as Pappus Chain.)

3.1 Steiner Chains with Nested Base Circles

This case is the most important and applicable and has been studied extensively. (See, Johnson 2007, Pedoe 1970, Roussos 2022, Yaglom 2002, etc.) We can call these Steiner Chains as **Classical Steiner Chains**. There are several published results that should be emphasized more and there are new ones that we are going to investigate and prove with regard to the classical Steiner chains. The whole study depends on inversion and the theorem that proves:

"For any two disjoint non-concentric circles there are inversions that map these two circles onto two concentric circles."

The center of such an inversion has only two choices, which are the **two Poncelet**[1] **points** of the two disjoint circles. The power of the inversion can be any. The **Poncelet points** are the points at which the smallest circle orthogonal to the two disjoint non-concentric circles intersects their center line. See **Figure 3.1**. The points P_1 and P_2 are the Poncelet points and the red circle is the **Poncelet circle**. The line HK is the radical axis of the two circles (O_1, R_1) and (O_2, R_2). The point M is the midpoint of O_1O_2. If the circles are nested, the situation is depicted in **Figures 3.2** and **3.3**.

(The reader must be familiar with the theory of inversion, the radical axis of two circles, and the Poncelet points, in order to study this chapter efficiently. For the convenience of the readers we have included this material in **Chapter 4**.)

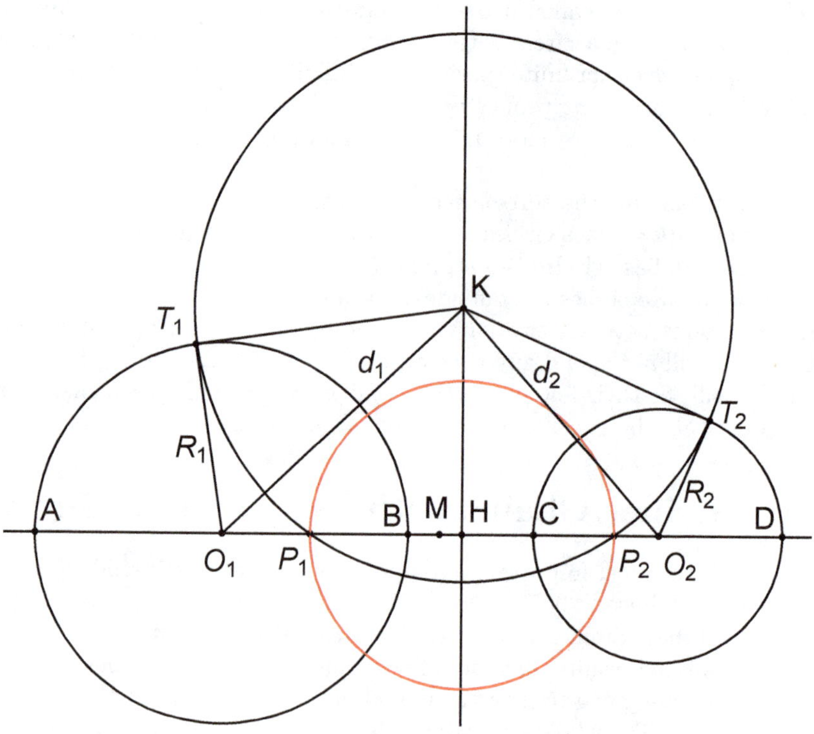

Figure 3.1: Poncelet Circle and Points

[1]Jean Victor Poncelet, French mathematician, 1788-1867.

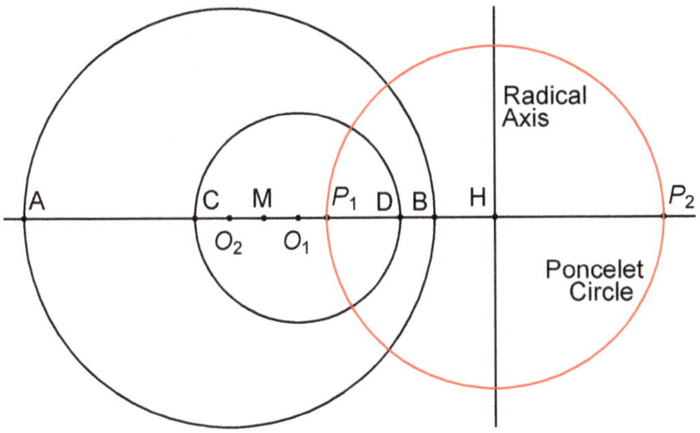

Figure 3.2: Poncelet Circle and Points

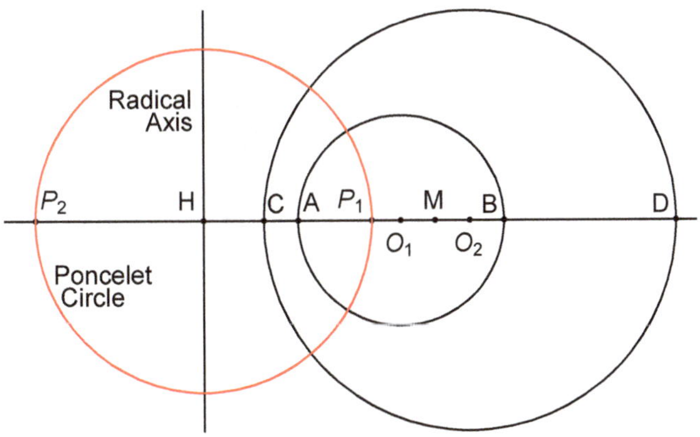

Figure 3.3: Poncelet Circle and Points

Before we continue with proving some results on the Steiner chains, we are going to state some general results. In **Figures 3.4** and **3.5**, we see two Steiner chains with **base circles** (O_1, R_1) and (O_2, R_2). The first one can continue to infinity without ever closing, that is, a circle becoming tangent to the first circle of the chain. We call such an infinite chain **open Steiner chain** (or **infinite or non-periodic Steiner chain**). The second chain closes after a finite number of circles, that is, a circle becomes tangent to the first circle of the chain. Such a chain can be continued periodically on the same circles. We call such a chain **closed Steiner chain** (or **finite**, or **periodic**). If the two base circles are concentric, we call the chain **canonical** (inside a canonical annulus).

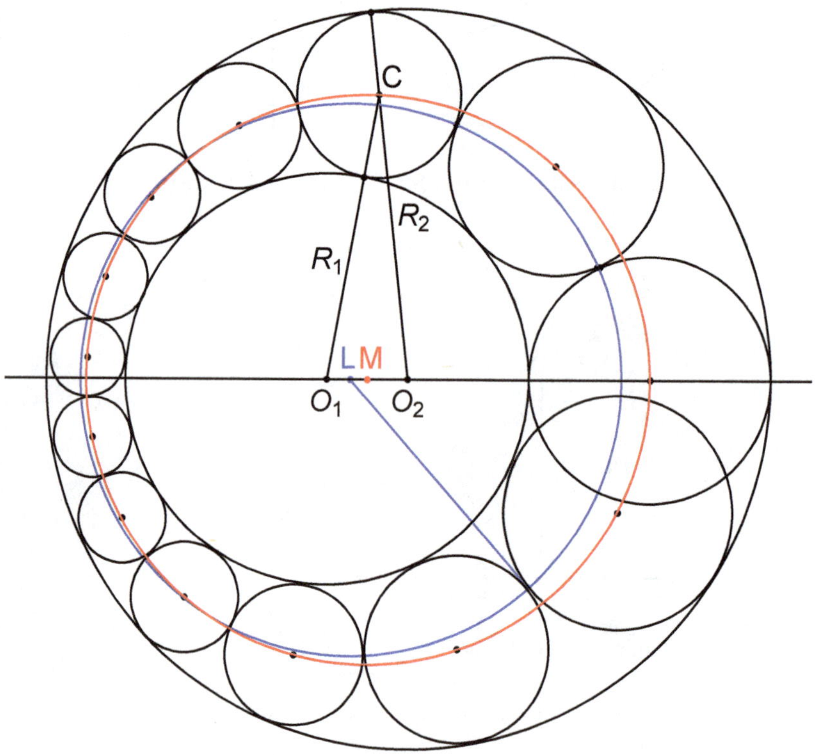

Figure 3.4: Open or Non-closing Steiner Chain

Remark: In both cases of **Figures 3.4** and **3.5**, it is a simple geometrical observation that the centers of the circles lie on an ellipse with foci the centers of the base circles O_1 and O_2 and so its center is the midpoint

M of O_1O_2. This so, because for any C center of a circle (C, r) of the chain (**Figure 3.4**), it is easily seen that

$$O_1C + O_2C = (R_1 + r) + (R_2 - r) = R_1 + R_2 \quad \text{constant.}$$

So, the major semi-axis of the ellipse is

$$M_{sa} = \frac{R_1 + R_2}{2}$$

and the minor semi-axis is

$$m_{sa} = \frac{\sqrt{(R_1 + R_2)^2 - (O_1O_2)^2}}{2}.$$

So, this ellipse is completely determined by the two base circles. If the circles are concentric, then $O_1 = O_2$, then the ellipse becomes a circle. This ellipse is the geometrical locus of the centers of all circles tangent to both base circles (O_1, R_1) and (O_2, R_2), and not just the circles of a Steiner chain. This result is also valid when $R_1 = 0$, that is, the inner base circle (O_1, R_1) is just a point O_1.

Using inversion, we prove that the points of tangency of the circles of the chain are on a circle. We are going to determine the center and the radius of this circle. But, before we do that we must state the famous:

Theorem 3.1.1 (Steiner's Porism) *For Steiner chains within nested base circles we have:*
If one chain of circles between two given nested circles closes (that is, it becomes periodic), then we have the following results:
(1) Every chain between the two given circles closes.
(2) So, there are infinitely many closed chains between the two given circles.
(3) Any circle tangent to the two given nested circles belongs to a closed chain.
(4) The number of circles in every closed chain is the same.
(5) Therefore: If one chain does not close, then no chain closes.

(The **proof** of this famous result can be found in the aforementioned bibliography and other books and in **Chapter 4**.)

Now, to determined the (blue) circle with center L and radius q that contains the points of tangency of the circles of a Steiner chain with nested base circles, we are going to study **Figure 3.6**. (It is a bit congested, but we will examine its features slowly and everything will fall into place.)

We consider two nested circles (O_1, R_1) and (O_2, R_2), such that $R_1 < R_2$ and $O_1O_2 < R_2 - R_1$. The small circle has diameter AB and the large CD

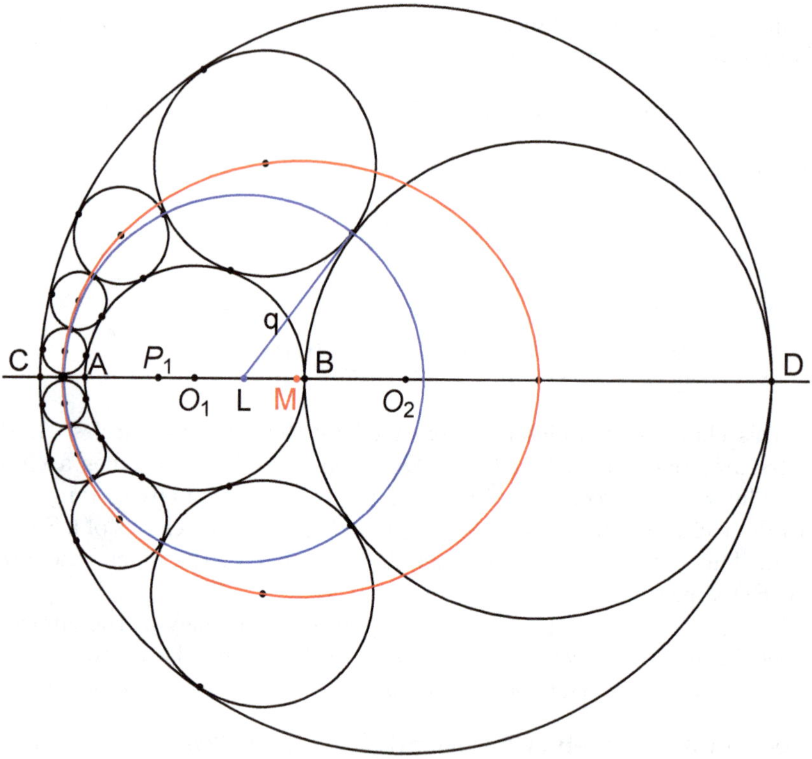

Figure 3.5: Closed Steiner Chain, n=9, m=1

on the center line O_1O_2.
We know that one of the Poncelet point, P_1 lt us say, lies inside the smaller circle.
We have drawn five circles of a chain with centers $C_1, \ldots, C_5$.
Let M be the midpoint of O_1O_2.
Let H be the foot of the radical axis of the two circles (O_1, R_1) and (O_2, R_2) on the center line O_1O_2.
To cover generality, we will use oriented segments when they are marked on the same straight line. In **Figure 3.6**, the positive orientation on the center line O_1O_2 is from left to right.
We must keep track of positive and negative segments and multiplication or division by them.
The squares of segments do not need orientation.
In our notation AB may mean the straight segment or line AB and / or

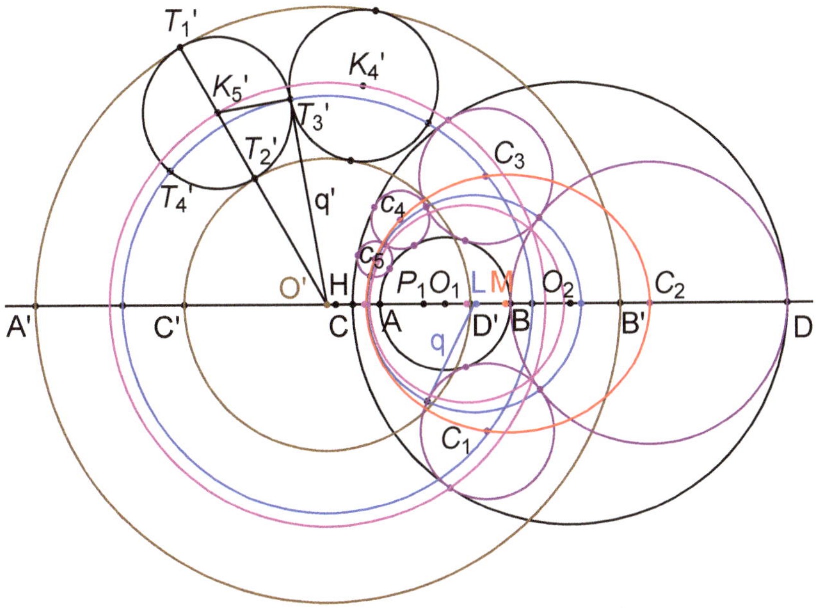

Figure 3.6: Inversion of a Steiner Chain to Determine the Circle (L, q)

the length of the segment AB, depending on the context.

From the study of radical axis of two circles, we have the following relations with the respected segments on the center line O_1O_2:

$$\overline{O_1M} = \overline{MO_2} \tag{3.1}$$

$$\overline{MH} = \frac{R_1^2 - R_2^2}{2\overline{O_1O_2}} \tag{3.2}$$

$$\overline{O_1H} = \frac{O_1O_2^2 + R_1^2 - R_2}{2\overline{O_1O_2}}, \tag{3.3}$$

$$\overline{O_2H} = \frac{-O_1O_2^2 + R_1^2 - R_2}{2\overline{O_1O_2}} \tag{3.4}$$

$$\sqrt{O_1H^2 - R_1^2} = |HP_1| = \frac{|P_1P_2|}{2} = |HP_2| = \sqrt{O_2H^2 - R_2^2} \tag{3.5}$$

The segments HP_1 and HP_2 have positive or negative orientation on the line O_1O_2, depending on the relative positions of the two initial circles (O_1, R_1) and (O_2, R_2), as shown in **Figures 3.1, 3.2, 3.3, 3.4, 3.5**, and

3.6. Since we know O_1O_2, R_1, and R_2, after we carry out some computations, we find the positions of the Poncelet points P_1 and P_2. In fact we have

$$\overline{P_1O_1} = \overline{P_1H} + \overline{HO_1}, \qquad \overline{P_1O_2} = \overline{P_1H} + \overline{HO_2} \tag{3.6}$$

$$\overline{P_2O_1} = \overline{P_2H} + \overline{HO_1}, \qquad \overline{P_2O_2} = \overline{P_2H} + \overline{HO_2} \tag{3.7}$$

{**Important Fact**: If $k = \overline{O_1O_2}$, $x = \overline{O_1P}$, where O_1 **is the origin**, P is the Poncelet point, and axis is $\overline{O_1O_2}$, then the solutions of the quadratic equation

$$kx^2 - \left(k^2 + R_1^2 - R_2^2\right) x + kR_1^2 = 0$$

(see, Chapter 4, or Roussos 2022, p. 351, for the derivation of this equation) give the position of the Poncelet points on the axis $\overline{O_1O_2}$. With $k = \overline{O_1O_2}$, the solutions are

$$\overline{O_1P} = x_\pm = \frac{k^2 + (R_1 + R_2)(R_1 - R_2)}{2k} \pm$$

$$\pm \frac{\sqrt{(k + R_1 + R_2)(k + R_1 - R_2)(k - R_1 + R_2)(k - R_1 - R_2)}}{2k}. \tag{3.8}$$

The two solutions coincide if and only if the circles are tangent $[k = \overline{O_1O_2} = \pm(R_1 - R_2)$, or $k = \overline{O_1O_2} = R_1 + R_2]$ and then, the Poncelet points coincide at the point of tangency. The two solutions are complex conjugate if and only if the circles intersect at two points ($|R_1 - R_2| < k < R_1 + R_2$), which are symmetrical in the axis $\overline{O_1O_2}$ and whose coordinates are the real and imaginary parts of the solutions. They coincide with the two common points of the circles. (The derivation of this formula is in **Chapter 4**, or Roussos 2022, pp. 351, 359.)

Similarly, if O_2 **is the origin** and $y = \overline{O_2P}$, we have the equation

$$ky^2 - \left(k^2 - R_1^2 + R_2^2\right) y + kR_2^2 = 0,$$

and now with $k = \overline{O_2O_1}$, the solutions are:

$$\overline{O_2P} = y_\pm = \frac{k^2 + (R_2 - R_1)(R_2 + R_1)}{2k} \pm$$

$$\pm \frac{\sqrt{(k + R_1 + R_2)(k + R_1 - R_2)(k - R_1 + R_2)(k - R_1 - R_2)}}{2k}.\} \tag{3.9}$$

The powers of the point P_1 with respect to the circles (O_1, R_1) and (O_2, R_2) are

$$\mathfrak{P}_1 = \overline{P_1A} \cdot \overline{P_1B} = P_1O_1^2 - R_1^2 \iff P_1O_1^2 = R_1^2 + \mathfrak{P}_1 \tag{3.10}$$

$$\mathfrak{P}_2 = \overline{P_1C} \cdot \overline{P_1D} = P_1O_2^2 - R_2^2 \iff P_1O_2^2 = R_2^2 + \mathfrak{P}_2 \tag{3.11}$$

Since the point P_1 lies inside both circles, these powers are negative. Also P_1 is on the same side (left or right) of both points O_1 and O_2. Now, these powers can be expressed by R_1, R_2 and O_1O_2, by using **equations 3.8** and **3.9**.

Now, we use an inversion $\mathfrak{I}$ *with center* P_1 *the Poncelet point inside both base circles*, and, without any loss, a **positive power** $c > 0$. This maps circles the (O_1, R_1) and (O_2, R_2) to the concentric circles (O', R_1') and (O', R_2') as in **Figure 3.6**. Here, $R_1' > R_2'$ and O' is the midpoint of $A'B'$ and $C'D'$, where A', B', C' and D' are the images of A, B, C and D under the inversion (diametrical points are mapped to diametrical points, under inversions).

As we know from the properties of inversion, the image circle of a given circle under an inversion is homothetic to the given circle, with center of homothety the center of the inversion and ratio of homothety the ratio of the power of the inversion divided by the power of the center of the inversion with respect to the given circle. This may be positive or negative. Therefore, for the radii of the image circles, we take the absolute value of this ratio and we have

$$R_1' = \left|\frac{c}{\mathfrak{P}_1}\right| R_1, \tag{3.12}$$

$$R_2' = \left|\frac{c}{\mathfrak{P}_2}\right| R_2 \tag{3.13}$$

$$\text{(here,}\quad c > 0,\ \ \mathfrak{P}_1 < 0,\ \ \mathfrak{P}_2 < 0)$$

Notice that, whereas we started with $R_1 < R_2$ and (O_1, R_1) is inside (O_2, R_2), now $R_1' > R_2'$ and (O', R_1') is outside (O', R_2'). In absolute value, we have

$$|\mathfrak{P}_1| < |\mathfrak{P}_2|.$$

In **Figure 3.6**, we have inverted the circles of the chain with centers C_4 and C_5. The inverted circles are equal, successively tangent and tangent to the image concentric circles (O', R_1') and (O', R_2'). From the figure, we can easily conclude that the points of tangency of the image circles are located on a circle with center O' and radius $O'T_3'$, equal to

$$q' = \sqrt{R_1' R_2'} = \frac{c\sqrt{R_1 R_2}}{\sqrt{|\mathfrak{P}_1 \mathfrak{P}_2|}} \tag{3.14}$$

By the same homotheties, we obtain

$$\overline{P_1O'} = \frac{c}{\mathfrak{P}_1}\,\overline{P_1O_1} \tag{3.15}$$

$$\overline{P_1O'} = \frac{c}{\mathfrak{P}_2}\,\overline{P_1O_2} \tag{3.16}$$

$$\left(\Longrightarrow \frac{\overline{P_1O_1}}{\mathfrak{P}_1} = \frac{\overline{P_1O_2}}{\mathfrak{P}_2}\right) \tag{3.17}$$

The relations **3.14** and **3.15** (or **3.16**) determine the circle (O', q') completely. The inverse image of this circle is the circle we want to determine, by finding its radius q and the position of its center L. To this end we need the power of P_1, *the Poncelet point inside both base circles*, with respect to this circle. By the equations **3.15** and **3.16**, $\mathfrak{P}'$ has the following two expressions

$$\mathfrak{P}' = P_1O'^2 - q'^2 = \frac{c^2}{\mathfrak{P}_1^2} P_1O_1^2 - \frac{c^2 R_1 R_2}{|\mathfrak{P}_1\mathfrak{P}_2|} = \frac{c^2}{\mathfrak{P}_2^2} P_1O_2^2 - \frac{c^2 R_1 R_2}{|\mathfrak{P}_1\mathfrak{P}_2|} \tag{3.18}$$

This power may be positive or negative, depending on the parameters involved.

Then, **the center L of the circle** (L, q) is located by

$$\overline{P_1L} = \frac{c}{\mathfrak{P}'}\overline{P_1O'} = \frac{c^2\overline{P_1O_1}}{\mathfrak{P}_1\mathfrak{P}'} = \frac{c^2\overline{P_1O_2}}{\mathfrak{P}_2\mathfrak{P}'} = \frac{\overline{P_1O_1}}{\mathfrak{P}_1\left(P_1O_1^2/\mathfrak{P}_1^2 - R_1R_2/|\mathfrak{P}_1\mathfrak{P}_2|\right)}$$

Therefore,

$$\text{if} \quad \alpha = \frac{\mathfrak{P}_1}{|\mathfrak{P}_1|},$$

$$\text{then} \quad \overline{P_1L} = \frac{|\mathfrak{P}_1\mathfrak{P}_2|\overline{P_1O_1}}{\alpha|\mathfrak{P}_2|P_1O_1^2 - \mathfrak{P}_1R_1R_2} = \frac{|\mathfrak{P}_1\mathfrak{P}_2|\overline{P_1O_1}}{\alpha|\mathfrak{P}_2|\left(\mathfrak{P}_1 + R_1^2\right) - \mathfrak{P}_1R_1R_2}, \tag{3.19}$$

and similarly,

$$\text{if} \quad \beta = \frac{\mathfrak{P}_2}{|\mathfrak{P}_2|},$$

$$\text{then} \quad \overline{P_1L} = \frac{|\mathfrak{P}_1\mathfrak{P}_2|\overline{P_1O_2}}{\beta|\mathfrak{P}_1|P_1O_2^2 - \mathfrak{P}_2R_1R_2} = \frac{|\mathfrak{P}_1\mathfrak{P}_2|\overline{P_1O_2}}{\beta|\mathfrak{P}_1|\left(\mathfrak{P}_2 + R_2^2\right) - \mathfrak{P}_2R_1R_2}. \tag{3.20}$$

In both expressions **3.19** and **3.20**, the $\overline{P_1O_1}$ and $\overline{P_1O_2}$ can be computed by **equations 3.15 and 3.16**, in terms of R_1, R_2 and O_1O_2, or use **3.10** and **3.11**.

By **3.14, the radius** q **of the circle** (L, q) is

$$q = \left|\frac{c}{\mathfrak{P}'}\right| q' = \frac{c^2}{|\mathfrak{P}'|} \cdot \frac{\sqrt{R_1R_2}}{\sqrt{|\mathfrak{P}_1\mathfrak{P}_2|}}.$$

Using relations **3.18** and **3.10**, we obtain the expression

$$q = \frac{|\mathfrak{P}_1|\sqrt{|\mathfrak{P}_1\mathfrak{P}_2|\sqrt{R_1R_2}}}{||\mathfrak{P}_2|P_1O_1^2 - |\mathfrak{P}_1|R_1R_2|} = \frac{|\mathfrak{P}_1|\sqrt{|\mathfrak{P}_1\mathfrak{P}_2|\sqrt{R_1R_2}}}{||\mathfrak{P}_2|\,(\mathfrak{P}_1 + R_1^2) - |\mathfrak{P}_1|R_1R_2|}, \tag{3.21}$$

and similarly by relation **3.11**, we obtain the expression

$$q = \frac{|\mathfrak{P}_2|\sqrt{|\mathfrak{P}_1\mathfrak{P}_2|\sqrt{R_1R_2}}}{||\mathfrak{P}_1|P_1O_2^2 - |\mathfrak{P}_2|R_1R_2|} = \frac{|\mathfrak{P}_2|\sqrt{|\mathfrak{P}_1\mathfrak{P}_2|\sqrt{R_1R_2}}}{||\mathfrak{P}_1|\,(\mathfrak{P}_2 + R_2^2) - |\mathfrak{P}_2|R_1R_2|} \tag{3.22}$$

Thus, we have completely determined the circle (L, q) that contains the points of tangency of the Steiner chain in which the base circles are nested. As we see, the formulae for the center L and the radius q do not depend on power of the inversion c but only on the base circles, a fact naturally expected. (In the formulae, notice the absolute values.)

Remark: The circle (L, q) intersects the circles of the chain orthogonally. Therefore, all common tangents of all pairs of two successive circles of a Steiner chain with nested base circles, at the common points of tangency, are radial lines of the circle (L, q) and so concurrent at its center L.

Note: The inverse images of the circles of the chain are equal circles tangent to the concentric images of the base circles. Their centers are on a concentric circle (O', σ'), of radius

$$\sigma' = \frac{R_1' + R_2'}{2} = \frac{|c|}{2}\left(\frac{R_1}{|\mathfrak{P}_1|} + \frac{R_2}{|\mathfrak{P}_2|}\right) = \frac{|c|}{2} \cdot \frac{R_1|\mathfrak{P}_2| + R_2|\mathfrak{P}_1|}{|\mathfrak{P}_1||\mathfrak{P}_2|}.$$

As we did with the circle (L, q), working in a similar way, we can find the circle (Θ, σ), which is the inverse image of the circle (O', σ'). This circle is located between the two base circles and has center on the line O_1O_2. In **Figure 3.6**, we have inverted the circle (O', σ'), that contains the centers K_1', K_2', ... This is a new circle drawn with the same color (magenta). So, the inverses of the points K_1', K_2', ..., are on this circle. But the centers of the circles of the chain, K_1, K_2, ... , are on the drawn red ellipse. For all n, the points K_n' and K_n are homothetic but not inverse of each other. The homotheties are determined as we have explained before (see the second paragraph after **equation 3.11**, or the bibliography).

Exercise: Find the radius σ and the center Θ of the circle (Θ, σ). Hint: The circles (O', σ') and (Θ, σ) are inverses by the so far chosen inversion with center P_1 and power c. Then, the power of P_1 with respect to the circle (O', σ') is given by $\Delta = P_1O'^2 - \sigma'^2 = \dots$ (complete the computation), and then $\sigma = \left|\frac{c}{\Delta}\right| \sigma' = \dots$, and so on. We again have two expressions for σ.

Similarly we find two expressions of $\overline{P_1\Theta}$, working in the same way that we have found the $\overline{P_1L}$ above.

3.2 Closing S. C., with Nested Base Circles

We consider two nested circles (O_1, R_1) and (O_2, R_2), such that $R_1 < R_2$ and $0 < d := O_1O_2 < R_2 - R_1$. A Steiner chain that contains n circles closes (becomes periodic) for the first time after the points of tangency wrap around the base circles m times, if m and n are relatively prime positive integers, and the nested base circles, (O_1, R_1) and (O_2, R_2), satisfy the **Steiner formula**

$$(R_2 - R_1)^2 - d^2 = 4R_1R_2 \tan^2\left(\frac{m\pi}{n}\right), \quad \text{where} \quad d = O_1O_2. \tag{3.23}$$

Solving for $d = O_1O_2$, the **distance** of O_1 and O_2, this equation is viewed as

$$d^2 = (R_2 - R_1)^2 - 4R_1R_2 \tan^2\left(\frac{m\pi}{n}\right),$$

(and the necessary condition $0 < d < R_2 - R_1$ is fulfilled). We can also write this equation as

$$d^2 = (R_2 + R_1)^2 - 4R_1R_2 \sec^2\left(\frac{m\pi}{n}\right).$$

(The derivation of this formula is based on the fact that inversions preserve the double ratio, and, set-wise, the diameters of circles that pass through the center of inversion. See Pedoe 1970, Roussos 2022, etc. Also you can imitate the derivation of the analogous **formula 3.32**, in **Section 3.4**, or see **Chapter 4**.)

The positive integer numbers m and n are called the **characteristic numbers of the closed Steiner chain**, which without loss of generality, are relatively prime. Their ratio $\frac{m}{n}$ of the number of wraps m divided by the number of the circles of the chain n is called **the characteristic of the finite Steiner chain**. If $m = 1$, the chain is called **simple**. If $m > 1$ the chain is called **multicyclic**. In either case, $n \geq 2$. (If $n = 2$, then $m = 1$ and the inner circle degenerates to a point.) If m and n are not relatively prime, their greatest common divisor k is greater than one. Then, the positive integers $m_1 = \frac{m}{k}$ and $n_1 = \frac{n}{k}$ are relatively prime. In this case, n_1 circles close for the first time after they wrap around the base circles m_1 times and then they repeat on themselves periodically k times.

The chain in **Figure 3.5** is closed and simple. In **Figure 3.7** below, the chain contains $n = 10$ circles that wrap around the base circles $m = 3$ times. This chain is canonical since the two base circles are concentric. In canonical chains $d = 0$ and the Steiner formula simplifies a bit. An inversion of this chain whose center is different from the common center of

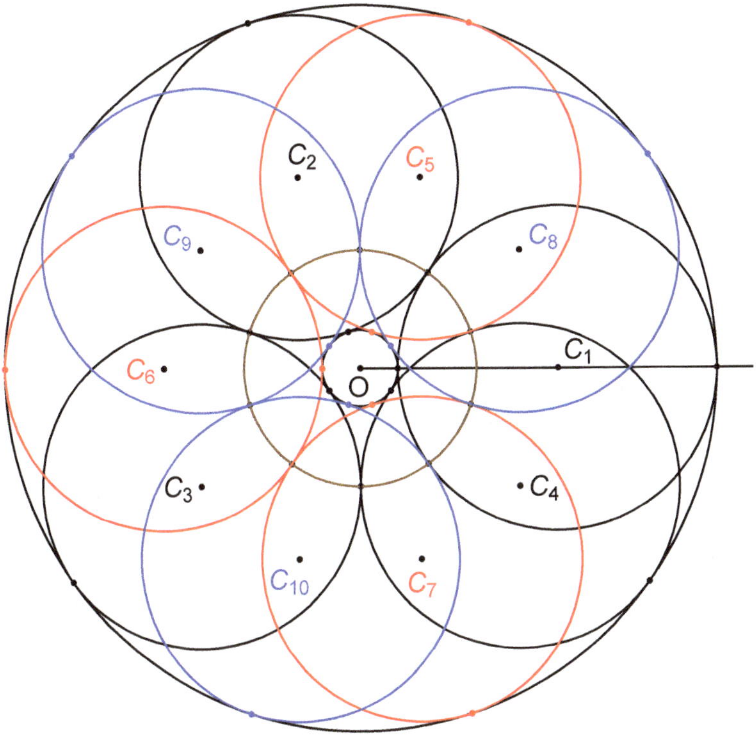

Figure 3.7: Multicyclic Canonical Steiner Chain

the concentric base circles, will change it to a Steiner chain between two non-concentric base circles.

In any closed chain the number of wraps m is determined in the following way. We count how many times the points of tangency to one of the base circles (any one of the two) with the circles of the chain trace the base circle. We start with a first circle of the chain, move along the base circle, successively from circle to circle of the chain, in the same direction, and finish at the first circle.

If $\frac{m}{n} = \frac{1}{2}$, then $\tan^2\left(\frac{\pi}{2}\right) = \infty$, and then for the **Steiner formula 3.23** to make any sense, we must have $R_1 = 0$, i.e., the small circle degenerates to a point. In any other case, we must have $\frac{m}{n} < \frac{1}{2}$. For example, we cannot have 10 circles wrapping around the base circles 7 times, or 3 circles 2 times, etc., as it is naturally seen. If $m = 1$ and $n = \infty$, then $R_1 - R_2 = d$, and so the circles are internally tangent and chain is a Pappus chain.

The equation of the Steiner formula is homogeneous of degree two, with resect to R, r and d. So, multiplying each of these variables by a (positive) factor the results and the characteristic $\frac{m}{n}$ of the chain remain the same. This, of course, was naturally expected.

We remark that it is straightforward to study the canonical Steiner chains and then transform the results obtained to ordinary chains, via inversions.

In a canonical chain, the base circles are concentric, and so $d = 0$. Then, we can solve for R, the radius of the outer circle, or for r, the radius of the inner circle, in terms of the remaining parameters, in order to construct a canonical closed chain of n circles.

Remarks: (1) If we allow the two circles to be internally tangent, and so

$$d = O_1O_2 = R_2 - R_1 > 0,$$

then we obtain Pappus chains as in **Figures 2.2** and **2.6**. In this case, **Formula 3.23** is valid, if $\frac{m\pi}{n} = 0$. This is so, if for any given $m = 1, 2, 3, \ldots$, the denominator $n = \infty$. This is exactly what happens with such Pappus chains.

(2) For concentric circles $d = 0$ and so with $R_2 > R_1$ by **Formula 3.23** we obtain

$$\tan\left(\frac{m\pi}{n}\right) = \frac{R_2 - R_1}{2\sqrt{R_1R_2}}.$$

Two practical ways to construct a simple canonical chain are the following:

(I) We make a canonical polygon with n sides. Then with centers its vertices and radii the half of its sides we draw the equal small circles. Then, with center the center of the canonical polygon and radii OD and OB, as in **Figure 3.8**, we draw the base circles.

(II) This way is base on the following three steps: Consider a circle $C[O, R]$ and two of its radii, Ox and Oy, making an angle $\measuredangle(Ox, Oy) = \alpha$, where $0 < \alpha \leq \pi$.

(a) Notice that the circle internally tangent to $C[O, R]$ and the two radii Ox and Oy, has center located on the bisector of the angle $\measuredangle(Ox, Oy)$ and its radius is

$$\rho = R \cdot \frac{\sin\left(\frac{\alpha}{2}\right)}{1 + \sin\left(\frac{\alpha}{2}\right)}.$$

Therefore, the (smallest) distance of O from such a circle is

$$r = R \cdot \frac{1 - \sin\left(\frac{\alpha}{2}\right)}{1 + \sin\left(\frac{\alpha}{2}\right)}.$$

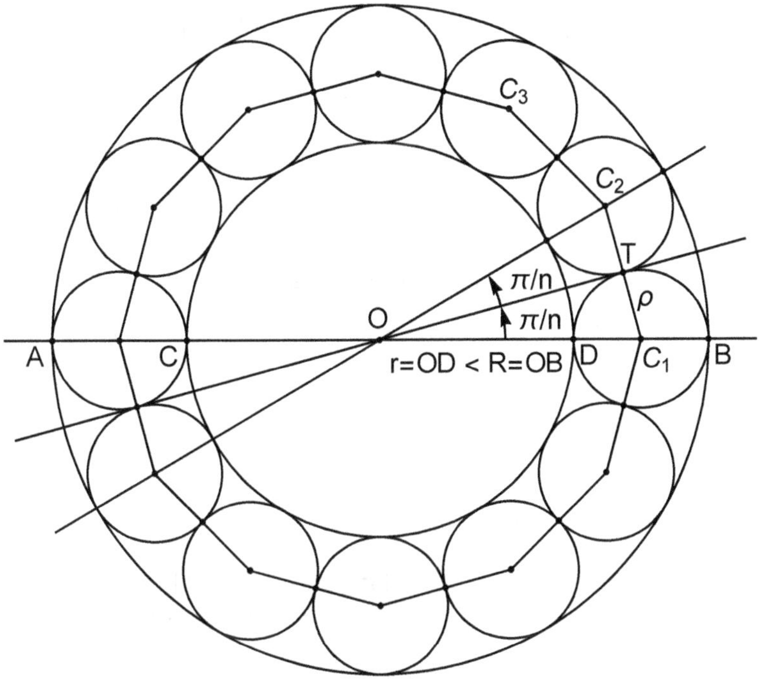

Figure 3.8: Canonical Simple Chain

(b) If $\alpha = \frac{2\pi}{n}$, where $n = 2,\ 3,\ 4\ \ldots$, then we can construct n equal circles, as in the previous step, which are successively externally tangent and the last circle is externally tangent to the first one.

(c) The circle $C[O, r]$ is externally tangent to all the n circles constructed in the previous step. These n circles form a simple, closed and canonical Steiner chain with base circles the two concentric circles $C[O, R]$ and $C[O, r]$.

Let us study a bit more a simple **canonical Steiner chain** with the help of **Figure 3.8**. We have imposed $R > r$. The radius of each circle of this canonical chain is

$$\rho := \frac{R - r}{2} = \frac{R + r}{2} \sin\left(\frac{\pi}{n}\right),$$

where $n = 2,\ 3,\ 4\ \ldots$. Therefore, for a canonical Steiner chain with characteristic $\frac{1}{n}$, the central angle is $360° \cdot \frac{1}{n} = 2\pi \cdot \frac{1}{n}$ and it holds

$$\sin\left(\frac{\pi}{n}\right) = \frac{R-r}{R+r}. \tag{3.24}$$

This implies, $R = r \cdot \frac{1+\sin\left(\frac{\pi}{n}\right)}{1-\sin\left(\frac{\pi}{n}\right)}$ and $r = R \cdot \frac{1-\sin\left(\frac{\pi}{n}\right)}{1+\sin\left(\frac{\pi}{n}\right)}$.

In a **multicyclic canonical Steiner chain** with characteristic $\frac{m}{n}$, as in **Figure 3.7**, the central angle is

$$\widehat{C_1OC_2} = 360° \cdot \frac{m}{n} = 2\pi \cdot \frac{m}{n}.$$

The radius of the outer base circle R, the radius of the inner base circle r, $(R > r)$, and

$$\rho = \frac{R-r}{2}$$

the radius of the equal circles of the chain are expressed by similar formulae. For example, **equation 3.24**, above, has general expression

$$\sin\left(\frac{m\pi}{n}\right) = \frac{R-r}{R+r} \tag{3.25}$$

and so on. As we explained earlier, the m and $n \geq 2$ can be considered relatively prime positive integers, without loss of generality. ($m = 1$ and $n = 2$ is a degenerate case.)

3.3 Steiner C's between a Circle and a Line

In **Figure 3.9** below, we see a chain of four circles tangent to a straight line l and to circle (O, R). Since the line l and the circle (O, R) are disjoint we also called such a chain, **Steiner chain.** It is a direct observation that in any such chain (finite or infinite) the centers of the circles of the chain are found on a parabola with focus the center O of the circle and directrix a line parallel to l at distance equal to the radius R and located on the side of l in which the circle (O, R) does not lie.

Here we will face two questions.

(A) How we obtain closed (finite, periodic) chains between a circle and a straight line.

(B) How we determine the circle that contains the points of tangency of the chain circles. Before we answer these questions, we must explain certain things concerning this case.

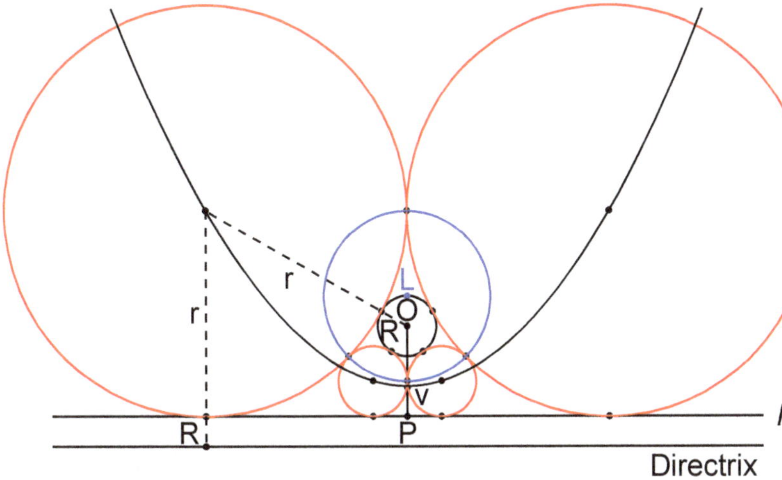

Figure 3.9: Line-Circle Finite Steiner Chain

In **Figure 3.10**, we describe the inversion(s) of a circle (O, R) and a line l that have no common point(s) into two concentric circles. The steps are the following and are easy to justify:
(i) From the center O of the circle we draw a line perpendicular to the given line, intersecting it at the point P.
(ii) From P we draw a tangent PT to the given circle.
(iii) We draw the circle (P, PT), which intersects the line OP at two points Q and $\overline{Q}$.
(iv) Any inversion with center Q, or $\overline{Q}$, will map the given circle and the line into two concentric circles.

So, we can use similar methods as in the case of two nested base circles to study these Steiner chains. In fact in **Figure 3.11**, we explain how to view this case as a limit of the ordinary Steiner chains, if we consider the line l as a circle of infinite radius and with center the infinity point in the direction of $\overrightarrow{PO}$.

We consider point K going to infinity along the line PO and in direction of $\overrightarrow{PO}$. For any such point we draw the circle (K, KP). As $K \longrightarrow \infty$, this circle approaches the line l. The foot H of the radical axis of (O, R) and (K, KP) approaches P and the Poncelet points P_K and $\overline{P_K}$ approach Q and $\overline{Q}$. In fact, we can say that the radical axis of a circle and a straight line is the straight line itself, the Poncelet points are the Q and $\overline{Q}$, and the Poncelet circle is the circle (P, PQ), which is the smallest circle orthogonal

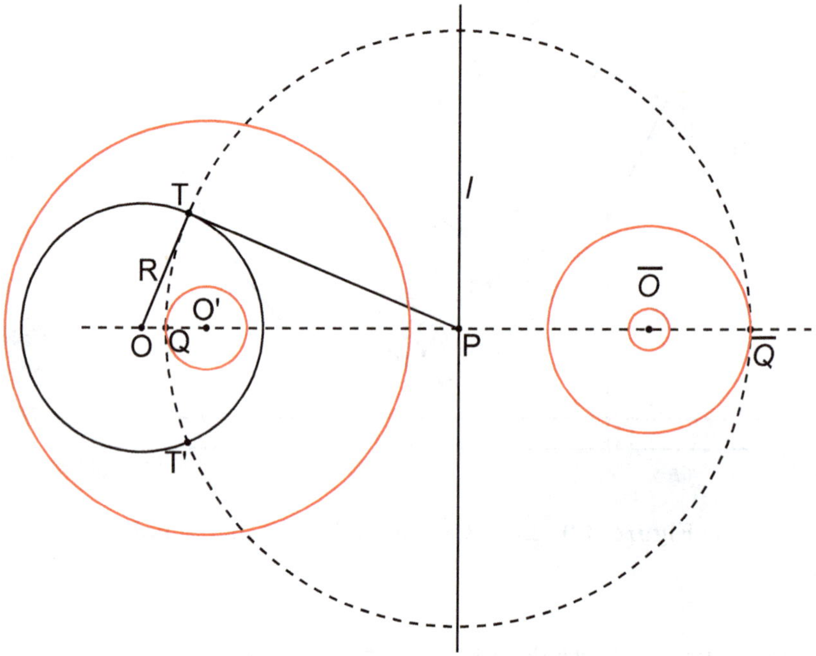

Figure 3.10: Inversion of Line and Circle into Concentric Circles

to both the circle (O, R) and the line l.

(A) Now we can answer what relation (O, R) and l must satisfy in order to obtain finite (closed, periodic) Steiner chain between them.

In the original **Steiner formula, 3.23**,

$$(R_2 - R_1)^2 - d^2 = 4R_1R_2\tan^2\left(\frac{m\pi}{n}\right),$$

$$\text{or} \quad \frac{(R_2 - R_1)^2 - d^2}{R_2} = 4R_1\tan^2\left(\frac{m\pi}{n}\right),$$

we let: $R_1 = R$, $d = d_K := KO$ (the distance of the two centers), and so we look at R_2 decomposed as

$$R_2 = KP = KO + OP = d_K + OP.$$

Then, we take limit, as $K \longrightarrow \infty \iff d_K \longrightarrow \infty$. That is,

$$\lim_{K\to\infty} \frac{(d_K + OP - R)^2 - d_K^2}{d_K + OP} = 4R\tan^2\left(\frac{m\pi}{n}\right).$$

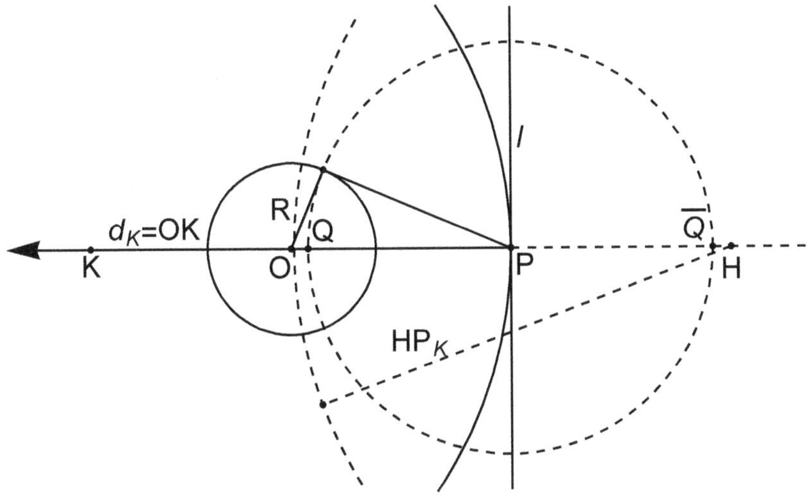

Figure 3.11: Limit of a Base Circle to a Line

This gives

$$OP - R = 2R\tan^2\left(\frac{m\pi}{n}\right),$$
$$\text{or}\quad OP = R\left[1 + 2\tan^2\left(\frac{m\pi}{n}\right)\right],$$

or the distance of the center of the given circle from the given line satisfies

$$OP = R\left[\tan^2\left(\frac{m\pi}{n}\right) + \sec^2\left(\frac{m\pi}{n}\right)\right] \tag{3.26}$$

This is the **new Steiner type formula** that (O, R) and l must satisfy, so that, a Steiner chain between them closes and has n circles that wrap around (O, R) m times, where m and n are positive integers such that $\frac{m}{n} \leq \frac{1}{2}$.

If $m = 1$ and $n = 2$, then $R = 0$ and the circle becomes a point. Non trivial chains exist, if $m = 1$ and $n \geq 3$. We also observe that if $m = 1$ (fixed in general) and $n \longrightarrow \infty$, then $OP \longrightarrow R$ and the Steiner Chain approaches a Pappus chain.

Remark: Using the limit process that we have used above, we can show that the radical axis of a circle and a straight line, as above, is the straight line and the Poncelet points are the Q and $\overline{Q}$. We need to rewrite the group of **equations 3.2, 3.3, 3.4**, etc., in terms of the radii and $\overline{OP}$ and then

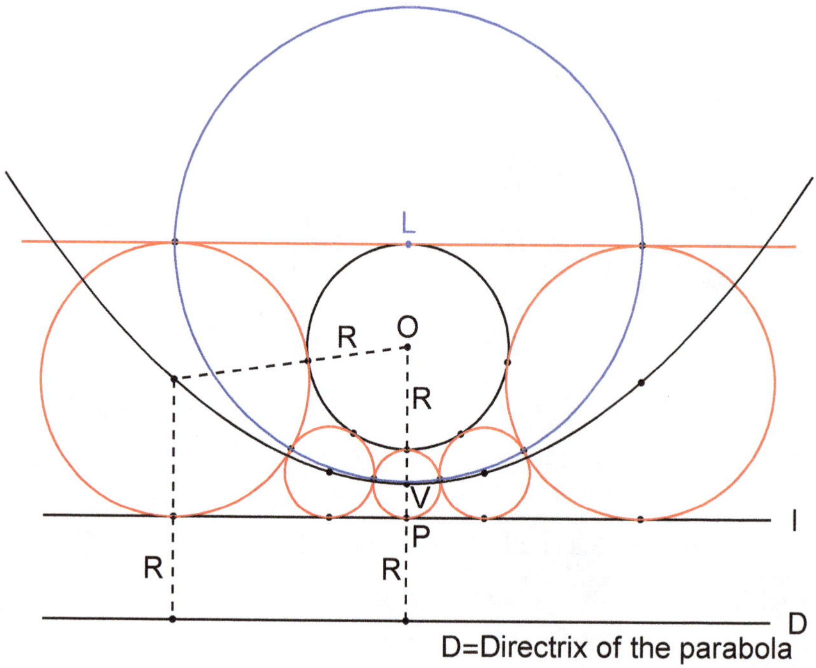

Figure 3.12: Line-Circle Steiner Chain

take the limit, as one radius is fixed and the other goes to infinity.

In **Figure 3.12**, we have $m = 1$ and $n = 6$. A special characteristic of this example is that one of the circle is a straight line tangent to the given circle (O, R) at the point of the intersection of the circle with its diameter OP at the side of O that the line l does not lie. This point is also the center of the circle that contains the points of tangency of the circles of the chain. (We will come back to this later because we need to also find the radius of this circle.) We must keep in mind that in this case of Steiner chains straight lines are also allowed and are considered as circles of infinite radius.This answers question (A).

Also, inverting this whole figure under an appropriate inversion, we create other periodic Steiner chains. **Figure 3.13** is an inversion of **Figure 3.12**, in which the base circles are not nested. Other inversions could have created chains between nested base circles. In **Figure 3.13** the centers of the circles of the chain are on a hyperbola (not parabola), but the points of tangency of the circles of the chain are on a circle whose center is on the center line of the two base circles.

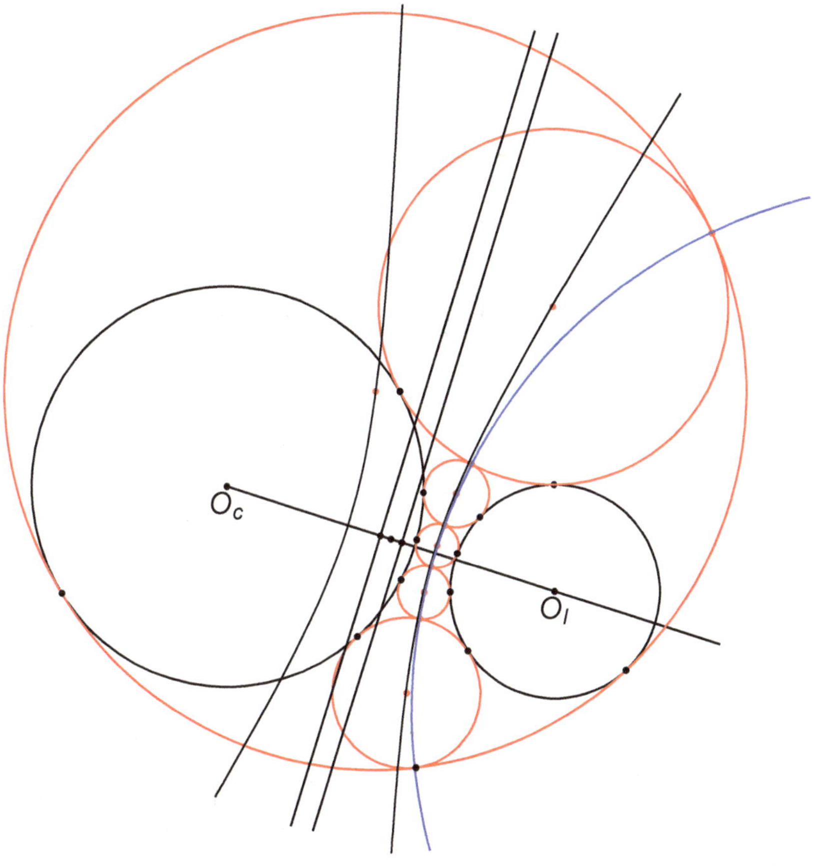

Figure 3.13: Inverse of Line-Circle Steiner Chain

(B) Now we want to determine the circle that contains the points of tangency of the chains as in **Figures 3.9** and **3.12**. We are going to refer to **Figure 3.14**. We consider a circle (O, R) and a line l at distance $s = OP > R$, and the inversion that maps the circle and the line to two concentric circles. The center of the inversion is Q (found earlier). We have that

$$QP^2 = OP^2 - R^2 \quad \Longleftrightarrow \quad QP = \sqrt{OP^2 - R^2},$$
$$\overline{OQ} = \overline{OP} + \overline{PQ},$$

and the power of Q with respect to the circle (O, R) is

$$\mathfrak{P} = OQ^2 - R^2.$$

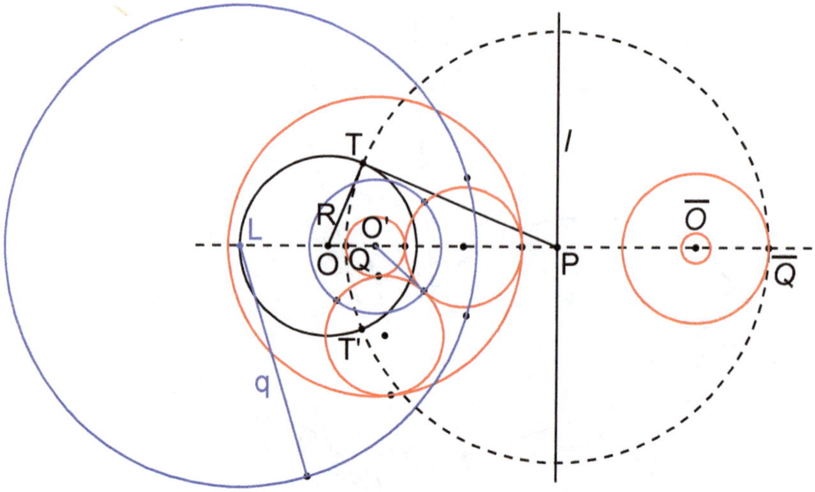

Figure 3.14: For the Study Line-Circle Steiner Chain

We work in the same way as we did in the previous section and we derive the formula for the radius q of this circle. We invert the line l and the circle (O, R), by an inversion with center Q and any power $c > 0$. The images are two concentric circles with center O'. We let P' be the inverse of P and the given straight line is mapped to the circle with diameter QP'.

Then we have the two equalities:

$$OQ = OP - QP,$$
$$QP^2 = OP^2 - R^2.$$

So,

$$R^2 = OP^2 - QP^2 = (OP + QP)(OP - QP) = (OP + QP)OQ =$$
$$(\overline{OP} + \overline{QP}) \cdot \overline{OQ}.$$

(This equality can be use as a resolvent relation in many computations left as **exercises**, below.)

The power of Q with respect to (Q, R) is:

$$\mathfrak{P} = QO^2 - R^2.$$

Also,

$$\overline{QP} \cdot \overline{QP'} = c \qquad \Longrightarrow$$
$$\overline{QP'} = \frac{c}{\overline{QP}},$$

and so,

$$R'_1 = \frac{c}{|\overline{QP}|}.$$

By the inverse of the straight line l, we have that O' is the midpoint of QP'. So,

$$\overline{QO'} = \frac{\overline{QP'}}{2} = \frac{c}{2\overline{QP}}.$$

Also,

$$\overline{QO'} = \frac{c}{\mathfrak{P}}\overline{QO}.$$

Hence we obtain:

$$\mathfrak{P} = 2\overline{QP} \cdot \overline{QO},$$
$$R'_2 = \frac{c}{|\mathfrak{P}|}R,$$
$$q' = \sqrt{R'_1 R'_2} = \sqrt{\frac{c^2 R}{2|\mathfrak{P}||\overline{QP}|}}.$$

The power of Q with respect to (O', R') is:

$$\mathfrak{P}' = QO'^2 - q'^2 = \frac{c^2}{4QP^2} - \frac{c^2 R}{2|\mathfrak{P}||\overline{QP}|}.$$

Now, working in a way similar to the derivation of q in previous section, we derive the formula

$$q = \left|\frac{c}{\mathfrak{P}'}\right| q' = \frac{2|QP|\sqrt{2|QP||\mathfrak{P}|R}}{||\mathfrak{P}| - 2R|QP||} \tag{3.27}$$

An equivalent formula can be found if we take the limit of **formulae 3.21** or **3.22** of the previous section. Thus,

$$K \longrightarrow \infty \quad \Longleftrightarrow \quad R_K = KQ + QP \longrightarrow \infty \quad \Longleftrightarrow \quad KQ \longrightarrow \infty.$$

Using **Figure 3.11**, we have

$$\mathfrak{P}_2 := \mathfrak{P}_K = KQ^2 - R_K^2 = KQ^2 - (KQ + QP)^2 =$$
$$-2KQ \cdot QP + QP^2$$

Thus, in this computation we must decompose R_K as $R_K = KQ + QP$. Now, if for example, we use **formulae 3.21** and take its limit, as $K \longrightarrow \infty$, we find another expression for q

$$q = \frac{|\mathfrak{P}|\sqrt{2|QP||\mathfrak{P}|R}}{|2|QP|(\mathfrak{P} + R^2) - |\mathfrak{P}|R|} \tag{3.28}$$

(Using all the relations among all segments involved and the power of Q, we can prove the equality of **Equations 3.27 and 3.28. Exercise**.)

Now, we must locate the center L of the circle (L, q). We will prove that L is the point of intersection of the given circle (O, R) with its diameter OP at the side of O that the line l does not lie.

Using the limit process, as stated above, and the **Formula 3.19**, we find

$$\overline{QL} = \frac{2|\mathfrak{P}||OP|\overline{QO}}{2\alpha(\mathfrak{P} + R^2)|QP| - \mathfrak{P}R}, \quad \text{where} \quad \alpha = \frac{|\mathfrak{P}|}{\mathfrak{P}}. \tag{3.29}$$

Then, by using all the relations involved among all segments and powers, we prove that the length of $\overline{OL} = \overline{OQ} + \overline{QL}$ (keep in mind the orientations of the segments) is equal to the radius R. This proves the claim stated above.

We can also derive an equivalent formula using the inversion with center Q and power $c > 0$. Then

$$\overline{QL} = \frac{c}{\mathfrak{P}'}\overline{QO'} = \frac{c^2}{\mathfrak{P}\mathfrak{P}'}\overline{QO}$$

and we derive the **simpler formula**

$$\overline{QL} = \frac{4QP^2 \cdot \overline{QO}}{\mathfrak{P} - 2\alpha R|QP|} = \frac{2\mathfrak{P}\overline{QP} \cdot \overline{QO}}{\mathfrak{P} - 2\alpha R|QP|}, \quad \text{where} \quad \alpha = \frac{|\mathfrak{P}|}{\mathfrak{P}}. \tag{3.30}$$

Exercise. Prove again that the length of $\overline{OL} = \overline{OQ} + \overline{QL}$ is R.

Remark: Again, the circle (L, q) intersects the circles of the chain orthogonally. Therefore, all common tangents of all pairs of successive circles of such a Steiner chain at the common points of tangency are radial lines of this circle and so concurrent at its center L.

3.4 S. C. with Non-nested Base Circles

We have already seen one such chain in **Figure 3.13**. In any such Steiner chain, the centers of the circles of the chain lie on a hyperbola with foci the centers O_1 and O_2 of the two base circles (O_1, R_1) and (O_2, R_2), and major semi-axis $\frac{|R_1 - R_2|}{2}$. If $R_1 = R_2$, the hyperbola degenerates to the straight line which is the mid-point perpendicular of the center segment O_1O_2. The points of tangency of the circles of the chain are on a circles whose center is on the center line O_1O_2 and can be determined by the methods we have exposed in the previous two sections. If $R_1 = R_2$, this circle (like the parabola) becomes the straight line which is the mid-point perpendicular of the center segment O_1O_2.

We would also like in this case, to find the Steiner type formula, analogous to **formula 3.23**, so that when the base circles are exterior of one another, a Steiner chain that contains n circles closes (or becomes periodic) after the circles wrap around the base circles m times. We look at the **Figure 3.15** and we rework the proof provided in Pedoe 1970, p.99, or Roussos 2022, p. 403. Also, see **Chapter 4**.

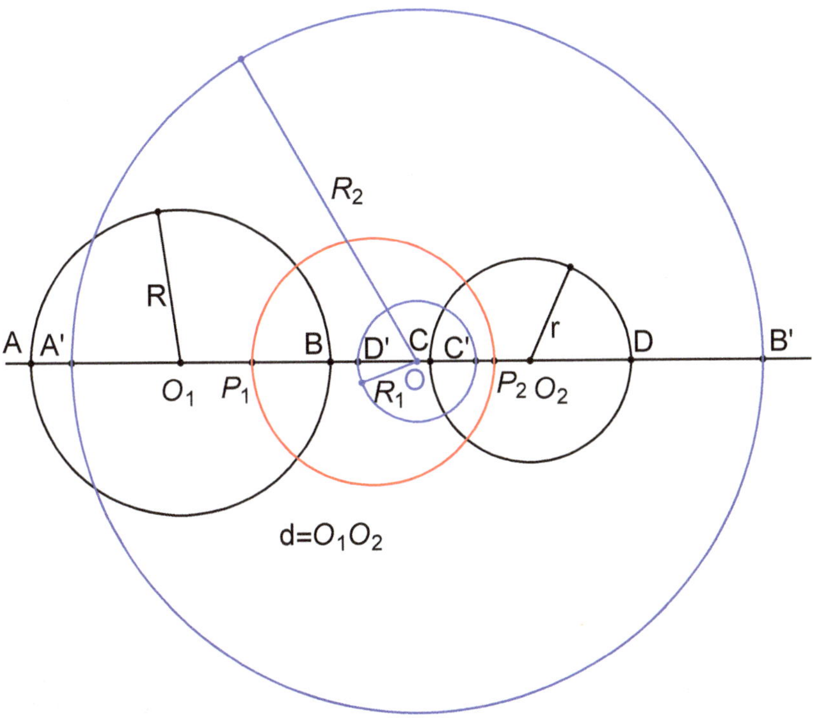

Figure 3.15: For the Relation of Finite Steiner Chain

The equality of the double ratios

$$[A,B;C,D] = [A',B';C',D'] \quad \Longrightarrow$$

$$\frac{\overline{AC}}{\overline{AD}} \cdot \frac{\overline{BD}}{\overline{BC}} = \frac{\overline{A'C'}}{\overline{A'D'}} \cdot \frac{\overline{B'D'}}{\overline{B'C'}}$$

gives

$$\frac{d^2-(R-r)^2}{d^2-(R+r)^2} = \frac{(R_2+R_1)^2}{(R_2-R_1)^2} \quad \Longleftrightarrow$$

$$d^2-(R-r)^2 = Rr\left(\frac{R_1}{R_2}+\frac{R_2}{R_1}+2\right).$$

With

$$\sin\left(\frac{m\pi}{n}\right) = \frac{R_2-R_1}{R_2+R_1} \quad \Longleftrightarrow$$

$$\frac{R_2}{R_1} = \frac{1+\sin\left(\frac{m\pi}{n}\right)}{1-\sin\left(\frac{m\pi}{n}\right)},$$

then, the new **Steiner type formula** obtained for this case is

$$d^2-(R-r)^2 = 4Rr\sec^2\left(\frac{m\pi}{n}\right) \tag{3.31}$$

$$\Longleftrightarrow$$

$$d^2 = (R+r)^2 + 4Rr\tan^2\left(\frac{m\pi}{n}\right), \tag{3.32}$$

$$\text{where} \quad R \geq r > 0, \quad \text{and} \quad d = O_1O_2 > R+r > 0.$$

Remark: Using this equation we can derive **equation 3.26** again, by taking limit as $R \longrightarrow \infty$ in a way analogous to the way we have used there.

Now, we would like to determine the circle (L,q) that contains the points of tangency of the circles of the chain. (The blue arc in **Figure 3.16**.) Working with **Figures 3.15** and **3.16** and the Poncelet point P_1, we again derive **formulae 3.19, 3.20, 3.21**, and **3.22**, in this situation. (Careful: Here, $\mathfrak{P}_1 < 0$ and $\mathfrak{P}_2 > 0$. Also, the centers of the circles of the chain are on the hyperbola stated above.)

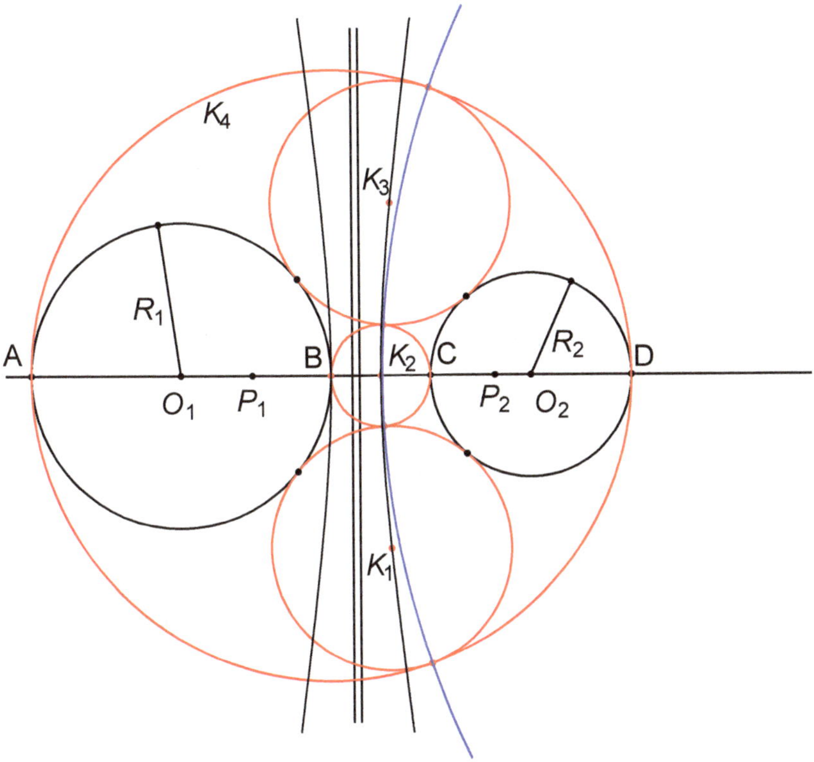

Figure 3.16: Hyperbola Containing the Centers of the Chain Circles

Remark: In a Steiner chain the centers of the circles of the chain are on a conic section: Circle, Ellipse, Parabola, Straight Line, Hyperbola, if the base circles respectively are: Concentric, nested but not concentric, one base circle is a straight line, both base circles are two different parallel straight lines, base circles are exterior to each other. (Study also **Section 3.8**.)

3.5 Conjugate Steiner Chains

We look at **Figure 3.16** and observe the following. The total number of circles is six. If we consider (O_1, R_1) and (O_2, R_2) as base circles, then the circles (K_1), (K_2), (K_3), and (K_4), is a Steiner chain between them of characteristic $\frac{1}{4}$. Again, if we consider (K_2) and (K_4) as base circles, then (O_1, R_1), (K_1), (O_2, R_2) and, (K_3), is a Steiner chain between them of characteristic $\frac{1}{4}$. Similarly, if we consider (K_1) and (K_3) as base circles, then (O_1, R_1), (K_2), (O_2, R_2) and, (K_4), is a Steiner chain between them of characteristic $\frac{1}{4}$. This is not an accident, but it happens whenever a Steiner chain has an even number of circles (C_1), (C_2), (C_3), ..., (C_{2k}).

In a chain with $2k$ even number of different circles, we call the circles C_1 and C_{k+1}, C_2 and C_{k+2}, and so on, **opposite circles or diametrical circles or antipodal circles**.

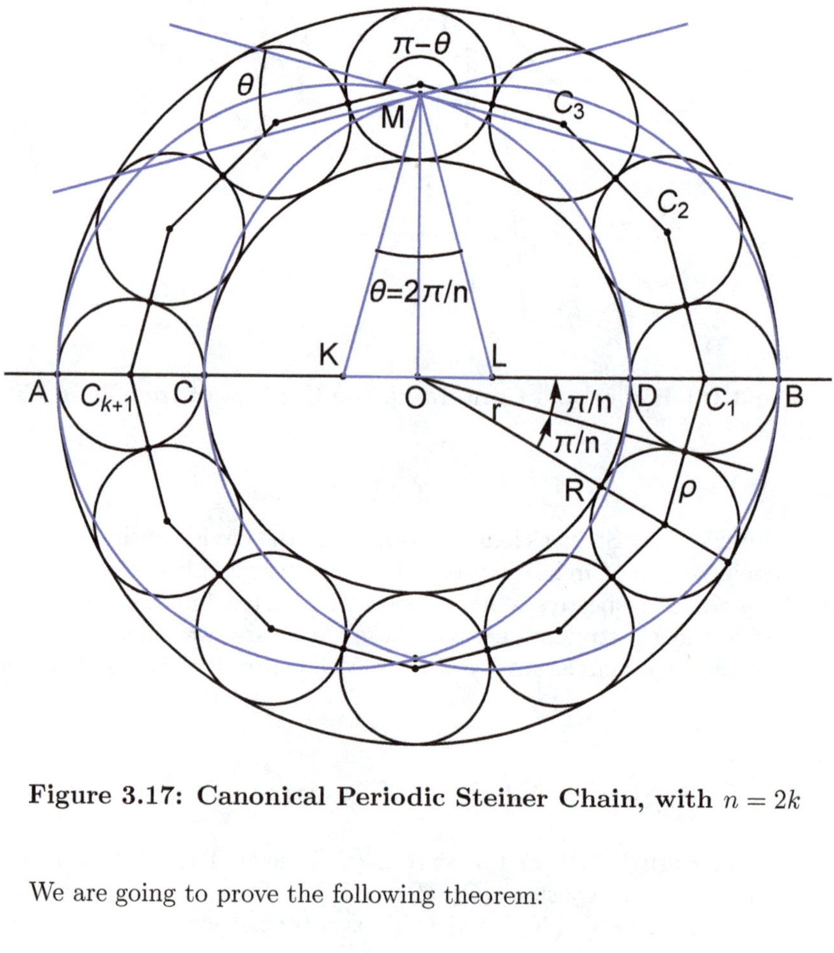

Figure 3.17: Canonical Periodic Steiner Chain, with $n = 2k$

We are going to prove the following theorem:

Theorem 3.5.1 *Let (C_1), (C_2), (C_3), ..., (C_n) be a Steiner chain with an even number of circles $n = 2k$, base circles (O_1, r) and (O_2, R), and characteristic $\frac{m}{n}$, where m and $n = 2k$ relatively prime positive integers (so m is odd). Then, with any opposite circles as base circles, we can fit another Steiner chain of characteristic $\frac{m'}{n'}$ such that*

$$\frac{m}{n} + \frac{m'}{n'} = \frac{1}{2}.$$

If n' is even, then we can find a chain that contains both base circles (O_1, r) and (O_2, R). (But, there are infinity many chains that do not contain either of the base circles.)

If n' is odd, then we can find a chain that contains only one of the base circles (O_1, r) and (O_2, R).[2] *(But, there are infinity many chains that do not contain any base circle.)*

This **Theorem** was proven by the great geometer **J. Steiner**, and surprised him so much that he considered it one of the most remarkable theorems in all geometry.

The new chains between two opposite circles are called **conjugate Steiner chains of the original chain.**

We have seen that any ordinary (finite) Steiner chain (simple or multicyclic) between two base circles can be brought to a canonical chain between two concentric base circles $C[O, R]$ and $C[O, r]$ with $R > r$, by means of an inversion. Therefore, we will study the result of **Theorem 3.5.1** with canonical chains.[3]

In any chain with base circles with diameters AB and CD, on the center line O_1O_2, as in **Figures 3.17, 3.18** and **3.19**, we can draw two circles with diameters AD and BC. These circles are internally tangent with the base circles in **Figures 3.17, 3.18** (same kind of tangency) and internally-externally **Figures 3.19** (alternate kind of tangency). (See also the **Examples** and the **Figures** at the end of this section.)

In **Figure 3.17**, we have a finite periodic chain with $m = 1$ and $n = 12$, the characteristic numbers of this chain. We could have any m and n with $n = 2k$ even number and $\frac{m}{n} < \frac{1}{2}$. So, with concentric base circles (O, R)

[2] Point missed in Johnson 2007, Theorem 164, page 115, in which it is stated that both base circles could belong to a conjugate chain. Yaglom 2009, expediently, says nothing to that effect. Studying **Figure 3.22** (in which we have $m = 1$, $n = 6$, $m' = 1$, and $n' = 3$), we directly see that we cannot find a conjugate chain that contains both base circles of the initial chain.

[3] In the conjugate chains some circles may become straight lines, circles of infinite radius. Two straight lines are tangent iff they are parallel.

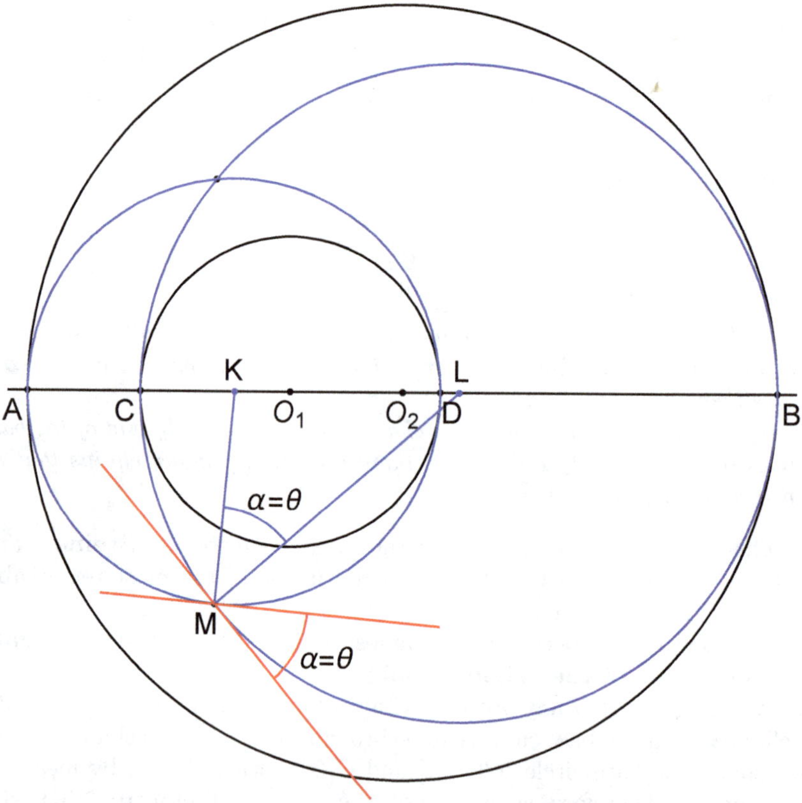

Figure 3.18: For the Condition of Periodic Steiner Chain

and (O, r), with $R > r$, any central angle between two rays from O and tangent to one of the equal circles of the chain, or through the centers of any two successive circles of the chain is equal to $\frac{m}{n} \cdot 2\pi$. We also have that the radius of any circle of the canonical chain is

$$\rho = \frac{R - r}{2}.$$

Now (in **Figure 3.17**), the circles with diameters AD and BC have centers the midpoints K and L of AD and BC, are equal, and

$$AD = CB = R + r,$$

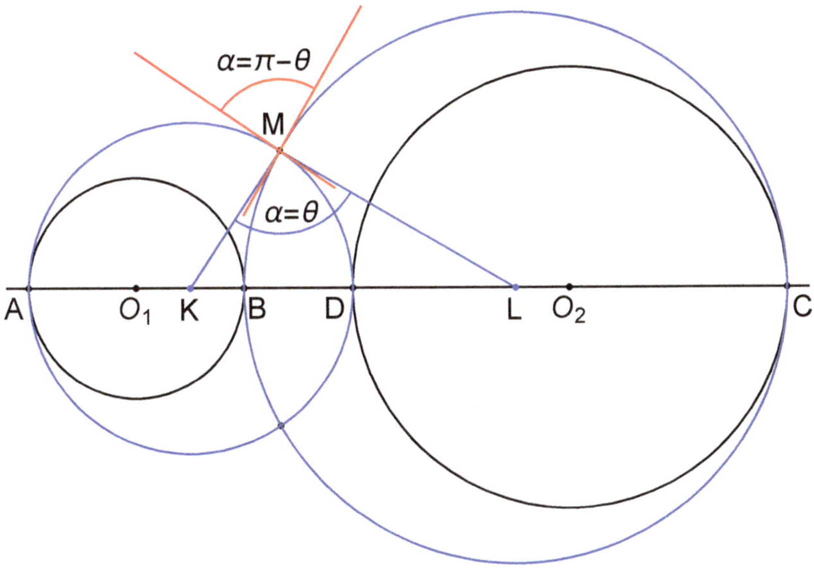

Figure 3.19: For the Condition of Periodic Steiner Chain

and so their radius is

$$s := \frac{1}{2}(R + r).$$

Let us call θ the angle of their intersection at one of their intersection point M. The distances of the centers K or L from the center O of the concentric base circles are equal

$$OK = OL = \frac{1}{2}(R - r)(= \rho), \quad \text{and so} \quad KL = R - r.$$

Also, MO is the height and median of the isosceles triangle KLM and has length

$$OM = \sqrt{\left(\frac{R+r}{2}\right)^2 + \left(\frac{R-r}{2}\right)^2} = \sqrt{Rr}.$$

We have put $\theta = K\hat{M}L$. This is equal to the angle between the corresponding tangents of the two circles (K, KM) and $(L, LM))$ at M. Consequently, by the law of cosines, we find

$$(R-r)^2 = 2\left(\frac{R+r}{2}\right)^2 - 2\left(\frac{R+r}{2}\right)^2 \cos(\theta) =$$
$$2\left(\frac{R+r}{2}\right)^2 [1-\cos(\theta)] = (R+r)^2 \sin^2\left(\frac{\theta}{2}\right),$$

from which we get,

$$\sin^2\left(\frac{\theta}{2}\right) = \frac{(R-r)^2}{(R+r)^2} \quad \Longrightarrow \quad 0 < \sin\left(\frac{\theta}{2}\right) = \frac{R-r}{R+r} < 1.$$

But, as we have seen in **equation 3.25**, this is also the $\sin\left(\frac{\pi m}{n}\right)$. Therefore,

$$\sin\left(\frac{\theta}{2}\right) = \sin\left(\frac{\pi m}{n}\right),$$

and with $0 < \theta < \pi$ and $\frac{m}{n} < \frac{1}{2}$, we have

$$\frac{\theta}{2} = \frac{\pi m}{n}.$$

Finally, for nested base circles, we have

$$\theta = 2\pi \frac{m}{n}.$$

Since equalities of non-oriented angles are preserve by inversions, we have discovered a **geometrical criterion** for a Steiner chain with base circles nested to close (be finite or periodic). This is:

*The angle α of the circles with diameters AB and CD, as in **Figures 3.17** and **3.18**, to be a rational fraction $\frac{m}{n}$ of 2π, i.e.,*

$$\alpha = \theta = 2\pi \frac{m}{n}.$$

For opposite circles, such as (C_1) and (C_{k+1}) in **Figure 3.17**, as base circles, the circles analogous two $\left(K, \frac{R+r}{2}\right)$ and $\left(L, \frac{R+r}{2}\right)$ for base circles the (O, R) and (O, r) remain the same. But, (C_1) and (C_{k+1}) are not nested and the angle corresponding to them at M is

$$\pi - \theta = \pi \frac{n-2m}{n} = 2\pi \frac{n-2m}{2n}.$$

This is also a rational fraction of 2π. We simplify this fraction to be

$$\frac{n-2m}{2n} = \frac{m'}{n'}$$

and, by the above criterion, we can create a finite (periodic) Steiner chain with characteristic $\dfrac{m'}{n'}$ and base circles two opposite circles.

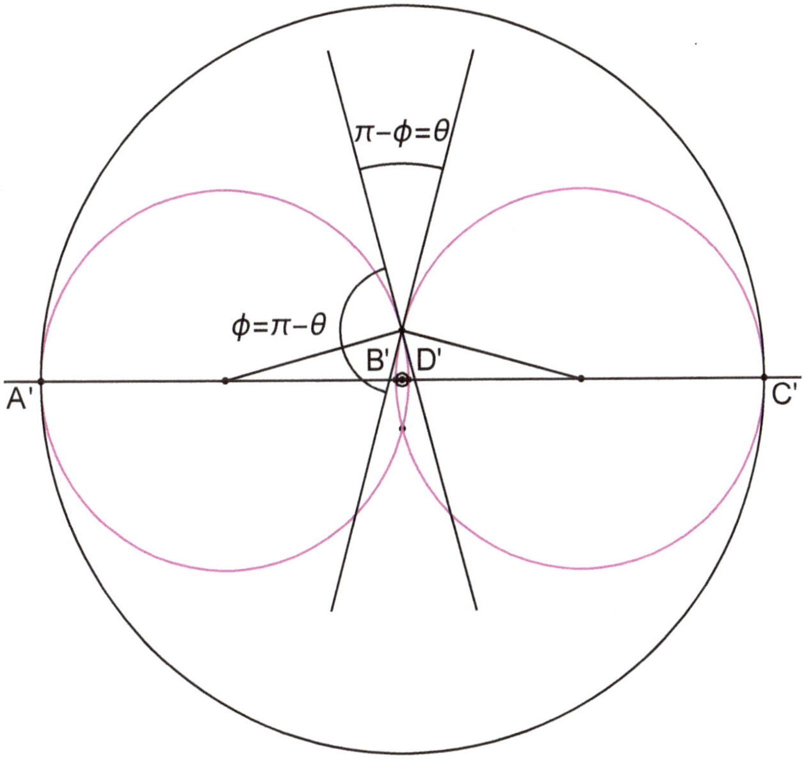

Figure 3.20: For the Condition of Periodic Steiner Chain

Inverting these apposite circles into concentric ones we obtain **Figure 3.20**. We see that one of the two opposite circles inverts to a small circle and the angle

$$\pi - \theta = 2\pi\frac{n-2m}{2n} = 2\pi\frac{m'}{n'}$$

transfers to the new figure indicated with the letter ϕ. Hence, we obtain a conjugate chain with base circle two opposite circles of the original chain with characteristic $\frac{m'}{n'}$.

But we see that,

$$2\pi\frac{m}{n} + 2\pi\frac{m'}{n'} = \pi.$$

Therefore,

$$\frac{m}{n} + \frac{m'}{n'} = \frac{1}{2}.$$

$$\text{(Or, quicker: } \frac{m}{n} + \frac{m'}{n'} = \frac{m}{n} + \frac{n-2m}{2n} = \frac{1}{2}.)$$

Example: If $\frac{m}{n} = \frac{1}{4}$ then $\frac{m'}{n'} = \frac{1}{4}$, as we saw that at the beginning of this section and **Figure 3.16**.

Similarly, if $\frac{m}{n} = \frac{1}{12}$ (**Figure 3.17**), then $\frac{m'}{n'} = \frac{5}{12}$.

If $\frac{m}{n} = \frac{3}{10}$, (**Figure 3.7**), then $\frac{m'}{n'} = \frac{1}{5}$, and so on.

If $m = 1$ and $n = 2$, then $\frac{1}{2} + \frac{1}{\infty} = \frac{1}{2}$, $r = 0$ and the conjugate chain degenerates to Pappus chain between two externally tangent circles.

With simplified characteristics to lower terms and n even, the relation

$$\frac{m}{n} + \frac{m'}{n'} = \frac{1}{2},$$

implies that m is odd. But, n' could be either odd or even. Also, both characteristics are less than $\frac{1}{2}$.

Constructing a conjugate chain is convenient to do it with canonical chains and then invert them to ordinary chains with nested or non-nested base circles. These constructions can show the validity of the assertions:

(a) If n' is even, then we can find a chain that contains both base circles (O_1, r) and (O_2, R).

(b) If n' is odd, then we can find a chain that contains one of base circles (O_1, r) and (O_2, R).

The canonical chains with n even are symmetrical with respect to the center of the concentric base circles and also symmetrical in the center line of two opposite circles, which is also a common diameter of the concentric base circles. These symmetries are also valid for the conjugate chain that are made on the same figure. Therefore, if n' is odd, it is impossible to keep both base circles in the conjugate chain and then fit $n' - 2$ (an odd number of) circles among them in a symmetrical way. But, for n' even, this is possible. Of course not every inversion preserves the validity of these assertions. All inversions but one will map a base circle in a way that its image does not belong to the conjugate chain. (See **Figures 3.22, 3.24, 3.25, 3.26** and **3.27** in the examples at end of this section that will make evident these assertions.)

The **existence of conjugate chains** between two opposite circles of an even Steiner chain can also be proven in a different way by using **equations 3.25** and **3.32** and with the help of **Figure 3.17**, as follows:

We observe that opposite circles exist, if the given Steiner chain has n an even number of different circles. Then for this chain we have, **equation 3.25**,

$$\sin\left(\frac{m\pi}{n}\right) = \frac{R-r}{R+r}. \tag{3.33}$$

The opposite circles are in the exterior of one another. So in this case, we must use **equation 3.32**, written as

$$d^2 - (R_1 - R_2)^2 = 4R_1R_2\sec^2\left(\frac{\theta}{2}\right),$$
$$\text{where} \quad R_2 \geq R_1 > 0,$$
$$\text{and} \quad d = O_1O_2 > R_1 + R_2 > 0.$$

We now check if this equation can be satisfied by some θ, which is a rational multiple of 2π for the opposite circles (C_1) and (C_{k+1}).

Looking at **Figure 3.17**, we see that for the opposite circles (C_1) and (C_{k+1}), we have:

$$R_1 = R_2 = \rho = \frac{R-r}{2}$$

and the distance of their centers is $d = R + r$. So, the equation becomes

$$\sec^2\left(\frac{\theta}{2}\right) = \left(\frac{R+r}{R-r}\right)^2 \quad \Longrightarrow \quad 0 < \cos\left(\frac{\theta}{2}\right) = \frac{R-r}{R+r}.$$

Therefore, by using **equation 3.33**, we have

$$\cos\left(\frac{\theta}{2}\right) = \sin\left(\frac{m\pi}{n}\right) = \cos\left(\frac{\pi}{2} - \frac{m\pi}{n}\right),$$

and we finally obtain

$$\theta = 2\pi\left(\frac{1}{2} - \frac{m}{n}\right) = 2\pi\frac{n-2m}{2n},$$

thus, proving the existence of the conjugate chain with characteristic

$$\frac{n-2m}{2n} = \frac{m'}{n'}, \qquad \text{(the same as before)}.$$

This result can also be verified with a usual non-canonical closed chain of n circles, with n even, as we see in **Figure 3.21**.

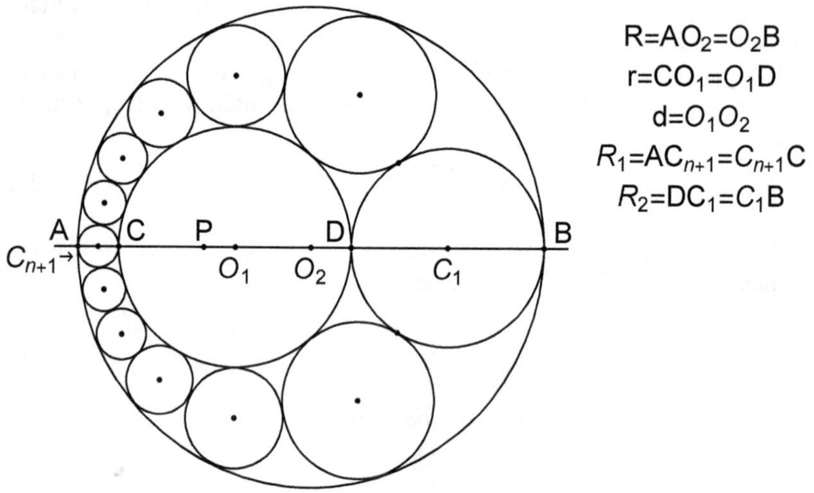

Figure 3.21: Non-Canonical Chain of Even Number of Circles

We have two non-concentric nested base circles (O_1, r) and (O_2, R) with $r < R$ and $d = O_1O_2$. We pick two opposite circles (C_1, R_2) and (C_{n+1}, R_1) with $R_1 < R_2$ with centers on the center line of the base circles, as in the figure. Let us put $\overline{d} = C_1C_{n+1}$. In this configuration, we directly find that

$$R_1 = \frac{R-r-d}{2}, \qquad R_2 = \frac{R-r+d}{2}$$

and so,

$$d = R_2 - R_1.$$

Also,

$$\overline{d} = R + r.$$

Now, we rewrite **equations 3.25** and **3.32** in this context and we get

$$(R-r)^2 - d^2 = Rr\tan^2\left(\frac{m\pi}{n}\right)$$

and

$$(R+r)^2 - d^2 = (R-r-d)(R-r+d)\sec^2\left(\frac{\theta}{2}\right).$$

Thus we get

$$\sin\left(\frac{m\pi}{n}\right) = \sqrt{\frac{(R-r)^2-d^2}{(R+r)^2-d^2}}$$

and

$$\cos\left(\frac{\theta}{2}\right) = \sqrt{\frac{(R-r)^2-d^2}{(R+r)^2-d^2}},$$

and so,

$$\cos\left(\frac{\theta}{2}\right) = \sin\left(\frac{m\pi}{n}\right).$$

As before, we again obtain

$$\theta = 2\pi\left(\frac{1}{2}-\frac{m}{n}\right) = 2\pi\frac{n-2m}{2n},$$

thus proving again the existence of the conjugate chain.

Exercise: Show that the same result is obtained, if we choose different opposite circles. (Of course, with a canonical chain this is obvious.)

Remark: *Conjugate Steiner chain of a closed chain with odd number of circles.* Since,

$$\frac{m}{n}+\frac{m'}{n'}=\frac{1}{2} \quad \Longleftrightarrow \quad 2(mn'+m'n)=nn',$$

if n' is odd, then n is even.

So, *given a chain with n'* **odd** *number of circles, the chain with n even number of circles and satisfying this equation is the conjugate chain of the given one.*

Examples

(I) We examine the **case** $m = 1$ **and** $n = 6$**, and so** $m' = 1$ **and** $n' = 3$**.** In **Figure 3.22**, we have a canonical chain with $m = 1$ and $n = 6$.

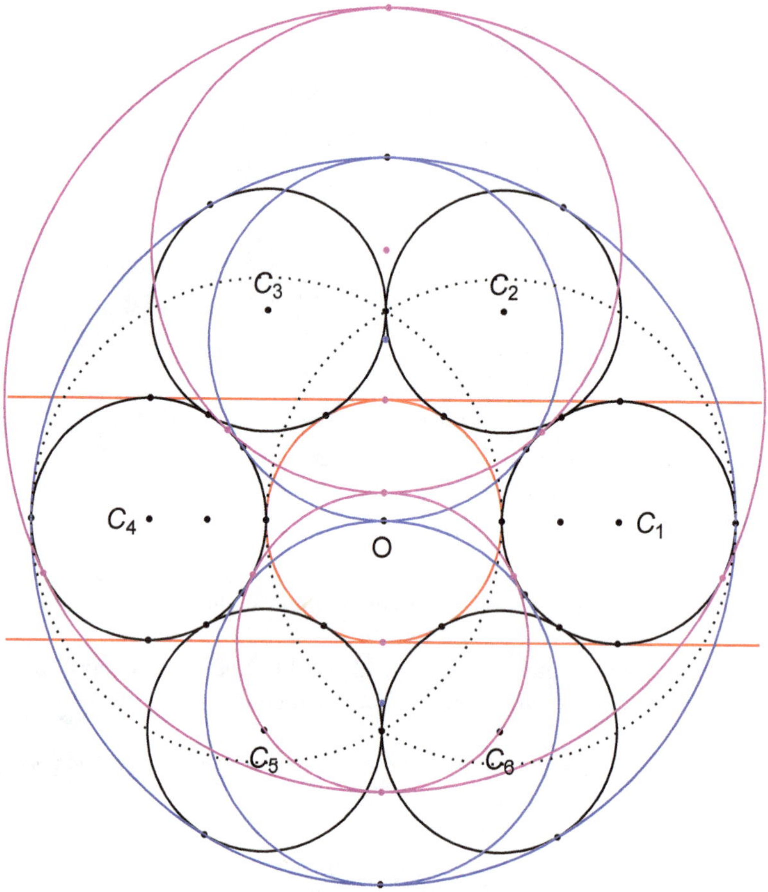

Figure 3.22: Canonical Chain and Three Conjugates with Base Circles (C_1) **and** (C_4)**, for** $m = 1$ **and** $n = 6$ **and so** $m' = 1$ **and** $n' = 3$

The base circles are the concentric circles with center O. The chain circles are (C_1), ..., (C_6). We choose opposite circles (C_1) and (C_4). (It is an easy exercise to prove that in this case the circles of the chain are equal to the small base circle.) Between them, we fairly easily construct three conjugate chains:

(a) The chain that consists of the central base circle and the two parallel lines. (Parallel lines are also admitted in the context of conjugate chains.)

(b) The chain that consists of the outer base circle and two other circles between this base circle and (C_1) and (C_4). The symmetry in the center line C_1C_4 shows that in this example it is not possible for both base circles to belong to a conjugate chain.

(c) The third chain (in purple) does not contain any of the two concentric base circles. (Its symmetric chain in the center line C_1C_2, not drawn in the figure, is also another such a chain.)

(d) Since in this case all circles of the chain are equal to the small base circle, at most one of the basic circles can be in the conjugate chain, never both of them.

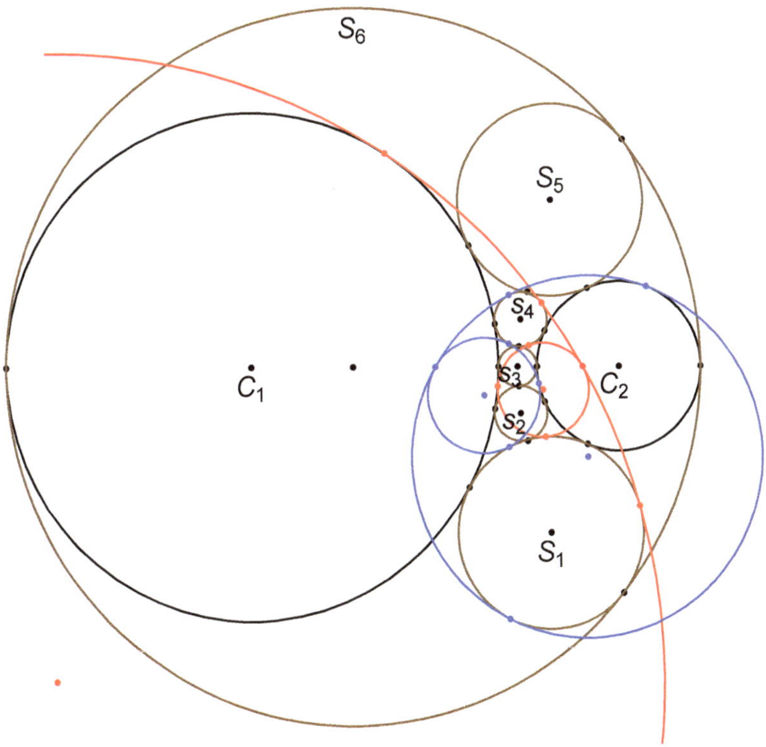

Figure 3.23: Two Inverted Chains, for $m = 1$ and $n = 6$, each containing only one of the base circles (C_1) and (C_2)

In **Figure 3.23** next, we have an ordinary chain with $m = 1$ and $n = 6$.

The base circles (C_1) and (C_2) are in the exterior of each other. The chain circles are (S_1), ..., (S_6). We choose opposite circles (S_1) and (S_4). Between them we have made two conjugate chains, one containing the red circles, and the other containing the blue circles, and each containing one of the base circles, only.

In **Figure 3.24**, we have an ordinary chain with $m = 1$ and $n = 6$. The base circles (C_1) and (C_2) are exterior of each other. The chain circles are (S_1), ..., (S_6). We choose opposite circles (S_1) and (S_4). Between them we have made two conjugate chains (one red, the other blue), none of which contains any of the base circles.

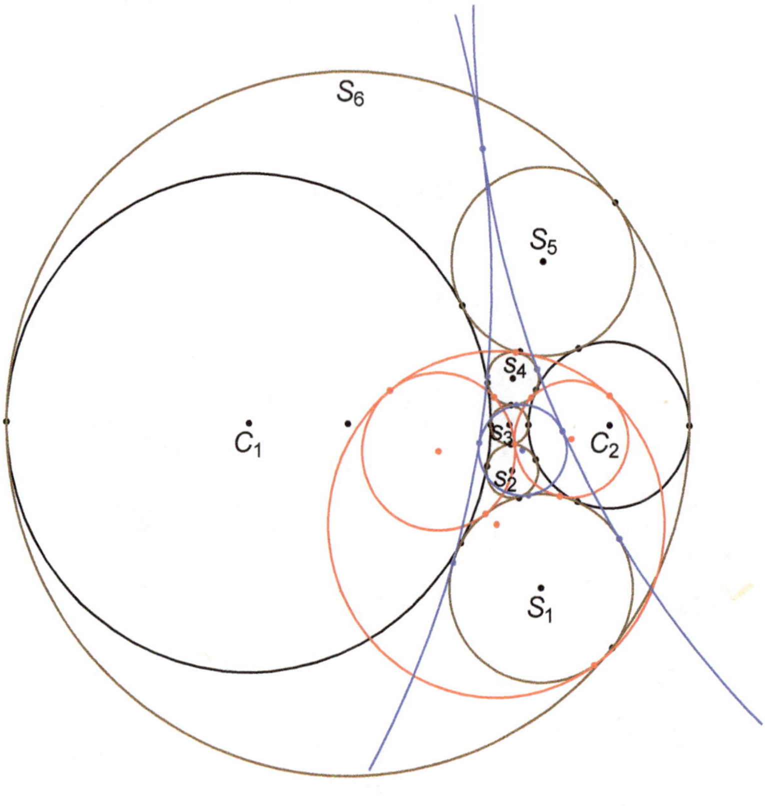

Figure 3.24: Two Inverted Chains, for $m = 1$ and $n = 6$, none of which contains any base circle (C_1) or (C_2)

Exercise: Make a classical Steiner Chain of four circles with nested base circles. The characteristic of this chain is $\frac{1}{4}$, and there are two pairs of opposite circles in this chain. Choosing any of these pairs, the conjugate chain consists of the other pair together with the two base circles, with characteristic $\frac{1}{4}$. (So, $\frac{1}{4} + \frac{1}{4} = \frac{1}{2}$, and there is no need to construct new circles for the conjugate chain.) We could call this chain as self-conjugate. This is the only self-conjugate chain.

(II) The chains in **Figures 3.23** and **3.24** are inversions of the chains in **Figure 3.22**. (Inversions could be found such that only one of the base circles belongs to the inverted chain.)

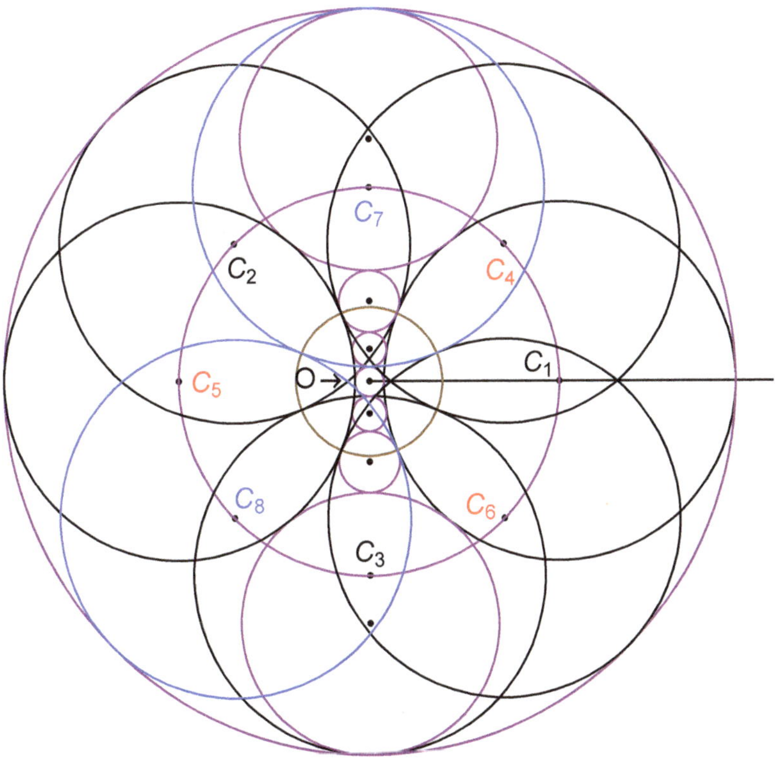

Figure 3.25: Canonical Chain and One Conjugate, for $m = 3$ and $n = 8$ and so $m' = 1$ and $n' = 8$, containing both concentric base circles

In **Figure 3.25** we have a canonical chain with $m = 3$ and $n = 8$. The base circles are the concentric circles with center O. The chain circles are (C_1), …, (C_8). We choose opposite circles (C_1) and (C_5). Between them, we have constructed a conjugate chain that contains both base circles ($m' = 1$ and $n' = 8$). The symmetry of the chain in the line C_1C_5 shows that this can be done in this case.

(III) In **Figure 3.26** below, we have a canonical chain with $m = 1$ and $n = 8$. The base circles are the concentric circles with center O. The chain circles are (C_1), …, (C_8). We choose the opposite circles (C_1) and (C_5).

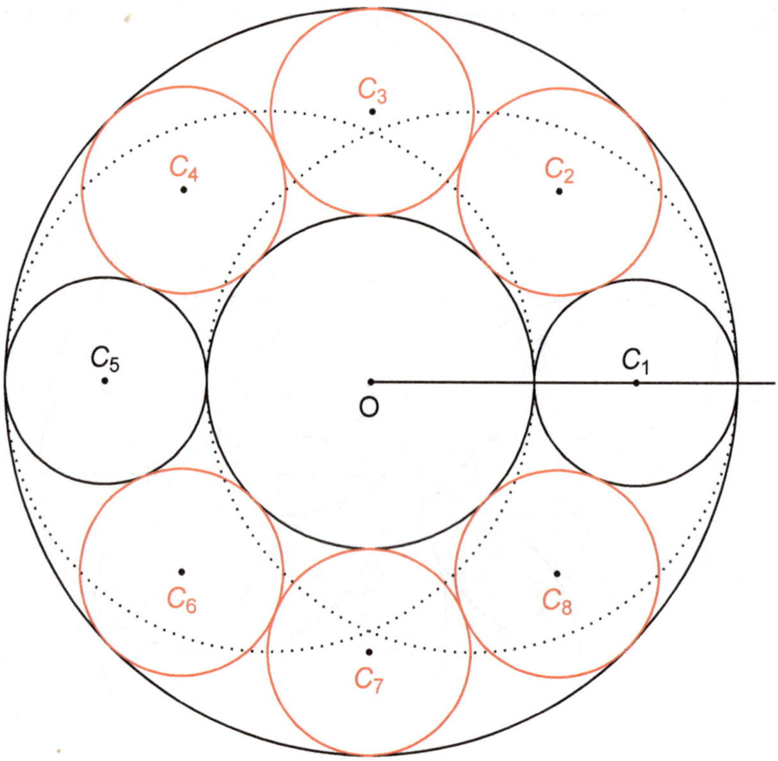

Figure 3.26: Canonical Chain, for $m = 1$ and $n = 8$ and so $m' = 3$ and $n' = 8$

In **Figure 3.27** below, we have constructed the conjugate chain ($m' = 3$ and $n' = 8$) of the above chain, with base circles (C_1) and (C_5). This conjugate chain contains both base circles of the original chain. The symmetry shows that this can be done in this case. As we see, this construction was more complicated. This wraps around (C_1) and or (C_5) 3 times. We have plot the points of tangency and indicated with numbers 1, 2, …, 8, and back to 1, how to follow the circles in order to see the three wraps.

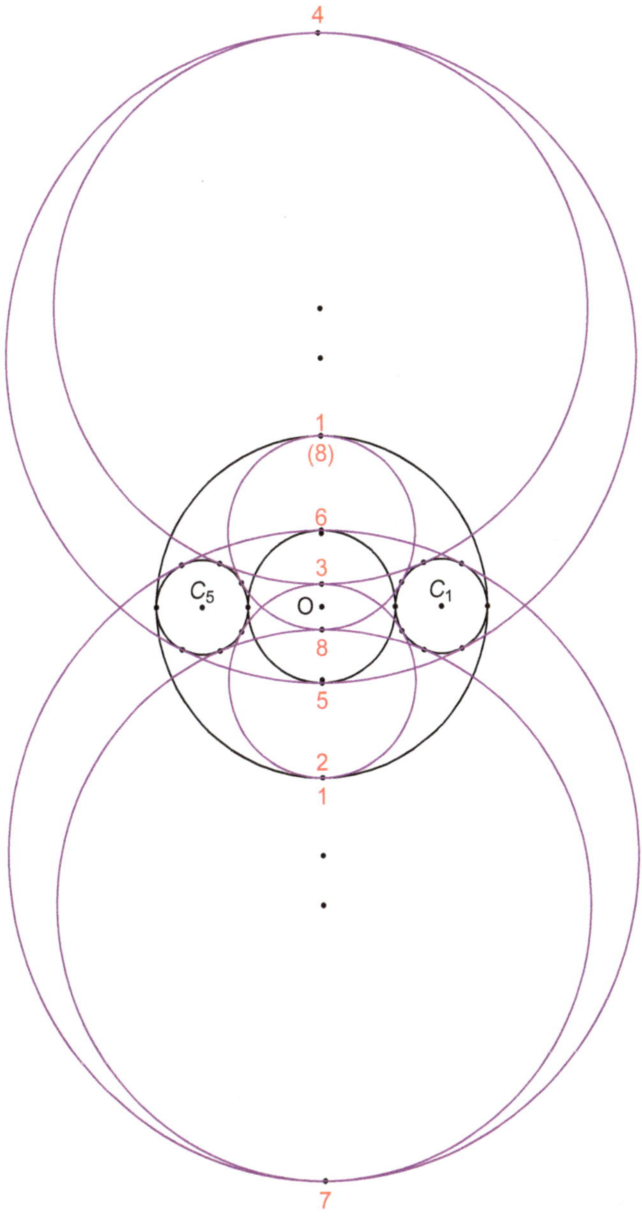

Figure 3.27: Conjugate Chain, for $m = 1$ and $n = 8$ and so $m' = 3$ and $n' = 8$, containing both concentric base circles

3.6 Radii & Centers for Classical S.C.

In this section, we find the **Archimedes type formula** that gives the radii and the centers of the circles of a Steiner Chain with nested base circles, see **Figure 3.28**. For this purpose, we need sufficient, not redundant and compatible data. These data are as follows:

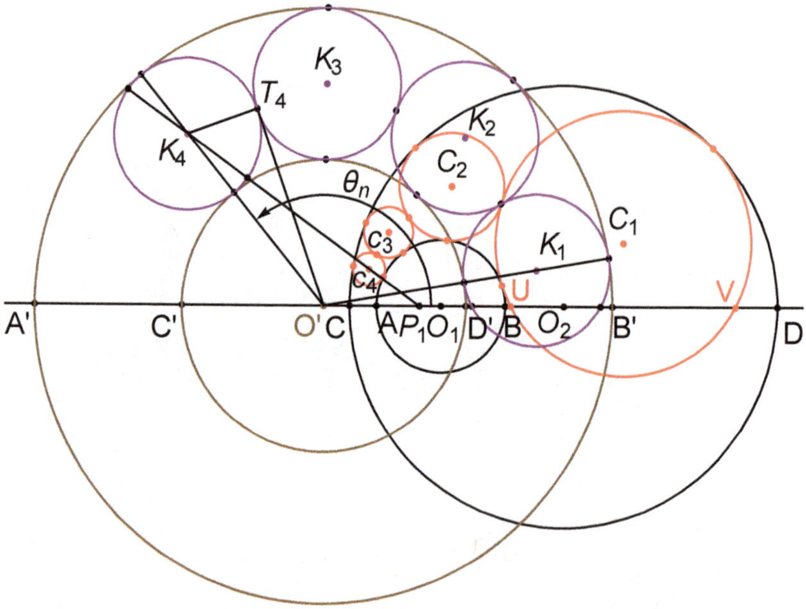

Figure 3.28: Steiner Chain with Nested Base Circles for computing the radii of the circles of the chain

We know the following: The two nested base circles (O_1, R_1) and (O_2, R_2) with $R_1 < R_2$, and $0 \leq d = O_1O_2 < R_2 - R_1$. We consider a Steiner chain of circles (C_1, r_1), (C_2, r_2), (C_3, r_3), ... , and so on. We always need a circle of the chain to be counted as the first one. So, we stipulate that the first circle of the chain (C_1, r_1) is such that, its center C_1 is above or on the center line O_1O_2 and intersects it at two points U and V. If $U = V$, then (C_1, r_1) is tangent to O_1O_2 and if C_1 is on O_1O_2, then O_1O_2 is a diameter and $U = B$ and $V = D$, in the figure. We need to know the radius r_1 of (C_1, r_1). We assume that these data are compatible in this problem.

Having these data at hand, we will find first the formula for r_n the radius of the nth of the chain. (A formula, analogous to Archimedes' formula for the Pappus chain, analogous to **Formulae 2.1, Theorem 2.1.3** and **2.3**) this will be achieved through the steps that follow. Second, we will find the center C_n.

We can determine the points C_1, U and V, by using well known **Stewart's Theorem** of classical plane geometry. We use **Figure 3.29**, and we see that if we know r_1 and we find points U and V on the center line O_1O_2, then the center C_1 is immediately determined. We will find O_2U and O_2V.

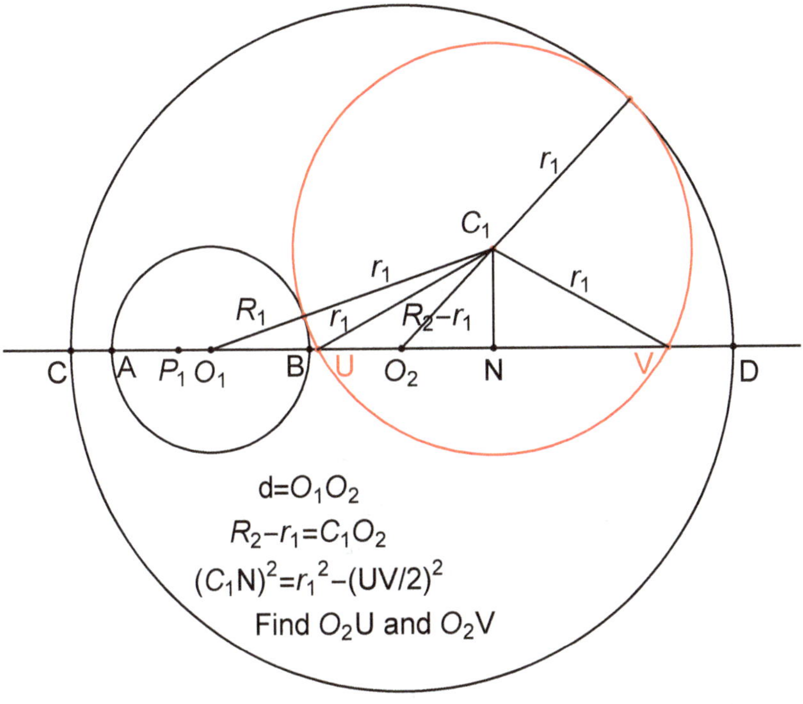

Figure 3.29: For the use of Stewart's Theorem

Given the tangencies of the three circles as in **Figure 3.29**, to find O_2V we solve the following quadratic equation, obtained from Stewart's Theorem applied to the triangle O_1C_1V, and we choose the positive solution.

$$(R_1 + r_1)^2 O_2V + r_1^2 d = (d + O_2V)[(R_2 - r_1)^2 + d\,O_2V]$$

and to find O_2U, we use the triangle $O_1C_1O_2$ and we solve

$$(R_1 + r_1)^2 O_2U + (R_2 - r_1)^2 (d - O_2U) = d[r_1^2 + (d - O_2U)O_2U]$$

Therefore, the points U and V and consequently C_1 (since we know r_1) are considered to be known. Then, known is also the midpoint N of the segment UV and the length of the segment C_1N, as shown in the figure.

Let $\mathfrak{P}_1'$ be the power of the Poncelet point P_1 of the base circles, which lies inside both base circles, with respect to the circle (C_1, r_1). Now, this is

$$\mathfrak{P}_1' = \overline{P_1U} \cdot \overline{P_1V} \tag{3.34}$$

This is positive in our configuration here, and also

$$\mathfrak{P}_1' = (P_1C_1)^2 - r_1^2.$$

So, we get

$$(P_1C_1)^2 = r_1^2 + \overline{P_1U} \cdot \overline{P_1V} \tag{3.35}$$

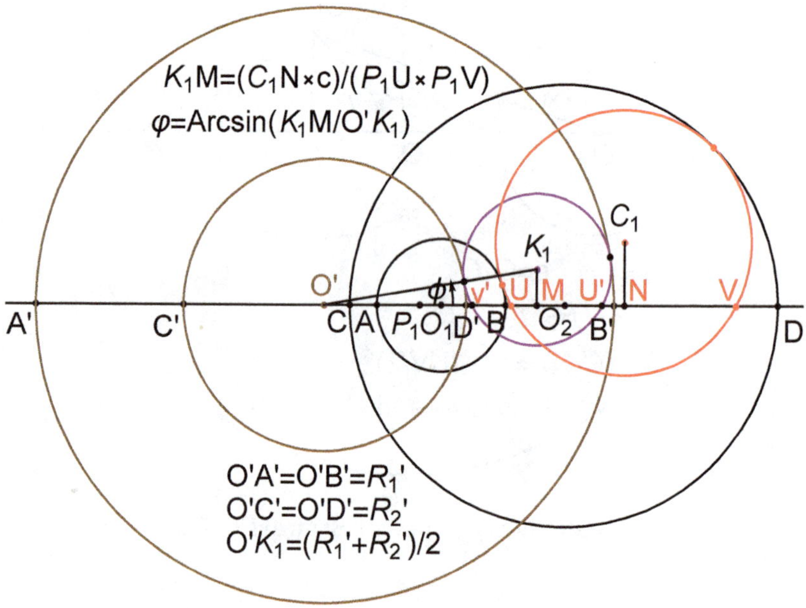

Figure 3.30: For computing Angle ϕ

In **Figure 3.30**, we have drawn the circle (K_1, ρ'), the inverse of (C_1, r_1) under the inversion with center P_1 and power $c > 0$ that maps the base circles to two concentric circles with center O'. We let M be the midpoint of $U'V'$, where U' and V' are the inverse points of U and V, under this inversion. (These are the points at which circle (K_1, ρ') intersects the line O_1O_2.) Now, we need to find the angle

$$\phi = \widehat{MO'K_1}$$

Since we have that (K_1, ρ') and (C_1, r_1) are homothetic, under the homothety with center P_1 and ratio $\frac{c}{\mathfrak{P}_1'}$, we find that

$$K_1M = \frac{c}{\mathfrak{P}_1'} C_1N$$

and

$$\phi = \arcsin\left(\frac{K_1M}{O'K1}\right).$$

But, as we have seen in **Section 3**, the radii of the concentric circles are

$$R_1' = R_1 \left|\frac{c}{\mathfrak{P}_1}\right|$$

and

$$R_2' = R_2 \left|\frac{c}{\mathfrak{P}_2}\right|,$$

where $\mathfrak{P}_1$ and $\mathfrak{P}_2$ are the known powers of P_1, with respect to the two base circles, respectively. So, with $c > 0$, we have

$$O'K_1 = \frac{R_1' + R_2'}{2} = \frac{c(R_1|\mathfrak{P}_2| + R_2|\mathfrak{P}_1|)}{2|\mathfrak{P}_1\mathfrak{P}_2|}.$$

Finally we obtain

$$\phi = \widehat{MO'K_1} = \arcsin\left(\frac{K_1M}{O'K_1}\right) = \arcsin\left(\frac{2|\mathfrak{P}_1\mathfrak{P}_2||C_1N|}{\mathfrak{P}_1'(R_1|\mathfrak{P}_2| + R_2|\mathfrak{P}_1|)}\right).$$

(Note: If in other configurations $\mathfrak{P}_1'$ is negative then we put it in absolute value in this formula.)

The radius ρ' of the circle (K_1, ρ'), and of all equal circles between the concentric circles, as we have in **Section 3**, is

$$\rho' = \frac{|R_1' - R_2'|}{2} = \frac{c(|R_1|\mathfrak{P}_2| - R_2|\mathfrak{P}_1||)}{2|\mathfrak{P}_1\mathfrak{P}_2|}$$

Now, if the inverse circles of the chain between the two concentric circles are (K_1, ρ'), (K_2, ρ'), (K_3, ρ'), ..., and so on, then each one of them occupies a central angle with vertex O' equal to

$$2\arcsin\left(\frac{\rho'}{O'K_1}\right) = 2\arcsin\left(\frac{||\mathfrak{P}_2|R_1 - |\mathfrak{P}_1|R_2|}{|\mathfrak{P}_2|R_1 + |\mathfrak{P}_1|R_2}\right).$$

Therefore, the central angle with vertex O', with initial side $O'D$ and terminal side OK_n, where K_n is the center of the nth circle (K_n, ρ'), is

$$\theta_n = \phi + 2(n-1)\arcsin\left(\frac{||\mathfrak{P}_2|R_1 - |\mathfrak{P}_1|R_2|}{|\mathfrak{P}_2|R_1 + |\mathfrak{P}_1|R_2}\right) =$$

$$\arcsin\left(\frac{2|\mathfrak{P}_1\mathfrak{P}_2||C_1N|}{\mathfrak{P}'_1(R_1|\mathfrak{P}_2| + R_2|\mathfrak{P}_1|)}\right) + 2(n-1)\arcsin\left(\frac{||\mathfrak{P}_2|R_1 - |\mathfrak{P}_1|R_2|}{|\mathfrak{P}_2|R_1 + |\mathfrak{P}_1|R_2}\right).$$

Then,

$$P_1K_n^2 = O'K_n^2 + O'P_1^2 - 2|O'K_n||O'P_1|\cos(\theta_n).$$

As we have done in **Section 3**,

$$\overline{O'P_1} = \frac{c}{\mathfrak{P}_1}\overline{P_1O_1} = \frac{c}{\mathfrak{P}_2}\overline{P_1O_2}.$$

Putting these together, after simplifying, we find that the power $\widetilde{\mathfrak{P}}_n$ of P_1 with respect to the circle (K_n, ρ'), is

$$\widetilde{\mathfrak{P}}_n = P_1K_n^2 - \rho'^2 =$$

$$c^2 \cdot \frac{|\mathfrak{P}_1|R_1R_2 + |\mathfrak{P}_2|P_1O_1^2 - (R_1|\mathfrak{P}_1| + R_2|\mathfrak{P}_1|)|P_1O_1|\cos(\theta_n)}{\mathfrak{P}_1^2|\mathfrak{P}_2|} =$$

$$c^2 \cdot \frac{|\mathfrak{P}_2|R_1R_2 + |\mathfrak{P}_1|P_1O_2^2 - (R_2|\mathfrak{P}_1| + R_1|\mathfrak{P}_2|)|P_1O_2|\cos(\theta_n)}{\mathfrak{P}_2^2|\mathfrak{P}_1|} \tag{3.36}$$

Then, we eventually obtain the formula for the radius r_n of the circle (C_n, r_n) of the Steiner chain

$$r_n = \frac{c}{\widetilde{\mathfrak{P}}_n}\rho' =$$

$$\frac{|\mathfrak{P}_1| \cdot ||\mathfrak{P}_2|R_1 - |\mathfrak{P}_1|R_2|}{2[|\mathfrak{P}_1|R_1R_2 + |\mathfrak{P}_2|P_1O_1^2 - (R_1|\mathfrak{P}_2| + R_2|\mathfrak{P}_1|)|P_1O_1|\cos(\theta_n)]} =$$

$$\frac{|\mathfrak{P}_2| \cdot ||\mathfrak{P}_1|R_2 - |\mathfrak{P}_2|R_1|}{2[|\mathfrak{P}_2|R_1R_2 + |\mathfrak{P}_1|P_1O_2^2 - (R_2|\mathfrak{P}_1| + R_1|\mathfrak{P}_2|)|P_1O_2|\cos(\theta_n)]} \tag{3.37}$$

(For a Pappus Chain the formula takes the undeterminate form $\frac{0}{0}$, since then P_1 is the point of tangency of the two base circles, and so $\mathfrak{P}_1 = 0 = \mathfrak{P}_2$.

Then, a limit process can derive **Formulae 2.1, Theorem 2.1.3** and **2.3**. Analogous work can be done for Steiner Chains with base circles exterior of each other, or if one base circle is a straight line. We leave it to the reader to imitate our steps here, adjust them to those configurations and derive the corresponding formulae.)

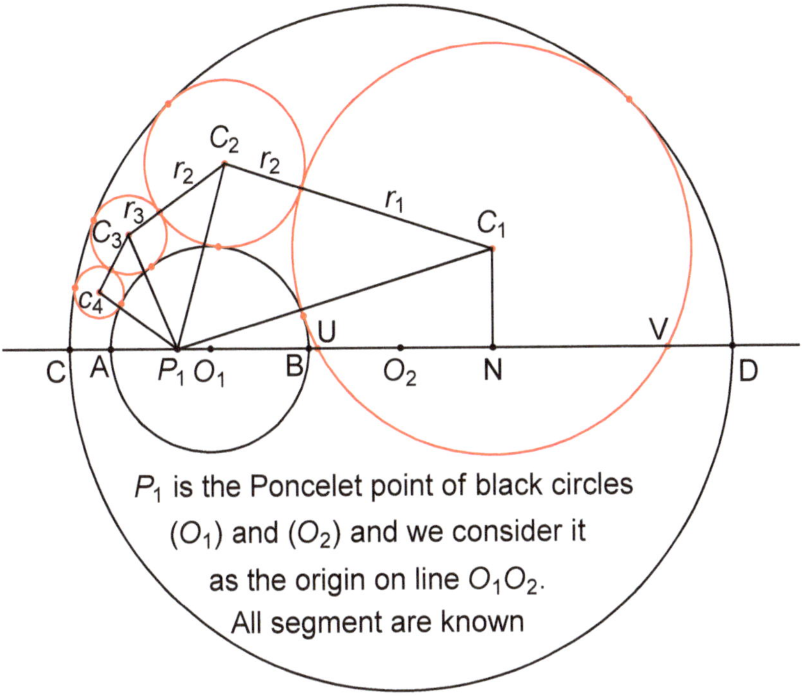

Figure 3.31: Steiner Chain with Nested Base Circles for computting the coordinates of the centers of the circles of the chain

Now, all the segments indicated in **Figure 3.31** can be computed by the previous results. Therefore, the centers of the circles of the chain can be determined. Namely, if we use as origin the Poncelet point P_1 which is inside both base circles in this configuration, and positive semi-axis P_1D, then we can compute the coordinates of each C_n, $n = 1, 2, 3, \ldots$.

We can solve every triangle $P_1C_nC_{n+1}$ because we know the three sides of it. We have $C_nC_{n+1} = r_n + r_{n+1}$ and we have already computed the radii r_n, $n = 1, 2, 3, \ldots$, from **equation 3.37**. The segment P_1C_1 is already known from **equation 3.35**. Next, we find each segment P_1C_n in the following way.

We have that

$$(P_1C_n)^2 = r_n^2 + \mathfrak{P}'_n$$

where $\mathfrak{P}'_n$ is the power of P_1 with respect to the circle (C_n, r_n). We know the power $\mathfrak{P}'_1$ of P_1 with respect to the circle (C_1, r_1), from **equation 3.34**.

Since the circles (C_n, r_n) and (K_n, ρ') are inverse of each other by the inversion with center P_1 and power some $c > 0$, we have

$$\mathfrak{P}'_n \cdot \widetilde{\mathfrak{P}}_n = c^2. \quad \text{So,} \qquad \mathfrak{P}'_n = \frac{c^2}{\widetilde{\mathfrak{P}}_n}$$

Then by **equation 3.36**, we get

$$\mathfrak{P}'_n = \frac{\mathfrak{P}_1^2|\mathfrak{P}_2|}{|\mathfrak{P}_1|R_1R_2 + |\mathfrak{P}_2|P_1O_1^2 - (R_1|\mathfrak{P}_1| + R_2|\mathfrak{P}_1|)|P_1O_1|\cos(\theta_n)} =$$

$$\frac{\mathfrak{P}_2^2|\mathfrak{P}_1|}{|\mathfrak{P}_2|R_1R_2 + |\mathfrak{P}_1|P_1O_2^2 - (R_2|\mathfrak{P}_1| + R_1|\mathfrak{P}_2|)|P_1O_2|\cos(\theta_n)}.$$

From **this equation** and **3.37**, we compute

$$(P_nC_n)^2 = r_n^2 + \mathfrak{P}'_n.$$

Now, every triangle $P_1C_nC_{n+1}$ can be solved, which in this case means that we can find its angles, since we know its sides. Then, the coordinates of the center $C_n = (x_n, y_n)$ in the system with origin P_1 and positive x-axis P_1D, are

$$x_n = P_1C_n\cos(\psi_n), \qquad y_n = P_1C_n\sin(\psi_n),$$

where $\psi_n = \widehat{DP_1C_n}$ (which can be computed).

Remarks: (1) For $n = 1, 2, 3, \ldots$ we have

$$\rho' = \frac{c}{\mathfrak{P}'_n}r_n, \qquad r_n = \frac{c}{\widetilde{\mathfrak{P}}_n}\rho', \qquad \text{and} \qquad \mathfrak{P}'_n\widetilde{\mathfrak{P}}_n = c^2,$$

Therefore,

$$r_n = \frac{\mathfrak{P}'_n}{\mathfrak{P}'_1}r_1, \qquad \text{and / or} \qquad r_n = \frac{\widetilde{\mathfrak{P}}_1}{\widetilde{\mathfrak{P}}_n}r_1.$$

Since r_1 is given from the beginning, we need to know the powers of P_1 with respect to the circles (C_n, r_n) or (K_n, ρ'), to compute r_n.

(2) Analogous work can be done for Steiner Chains with base circles exterior of each other, or if one base circle is a straight line. We leave it to the reader to imitate our steps here and find the formulae in those configurations.

3.7 Two Properties of the Closed S. C.

(1) One property of the closed (finite, periodic) Steiner chains is depicted in, **Figures 3.32** and **3.34**.

We make pairs of opposite points of tangency, then we make two groups of circles determined by these pairs and P_1, the Poncelet point inside the base circles. Then, all these circles of both groups are also concurrent at the other Poncelet point, P_2.

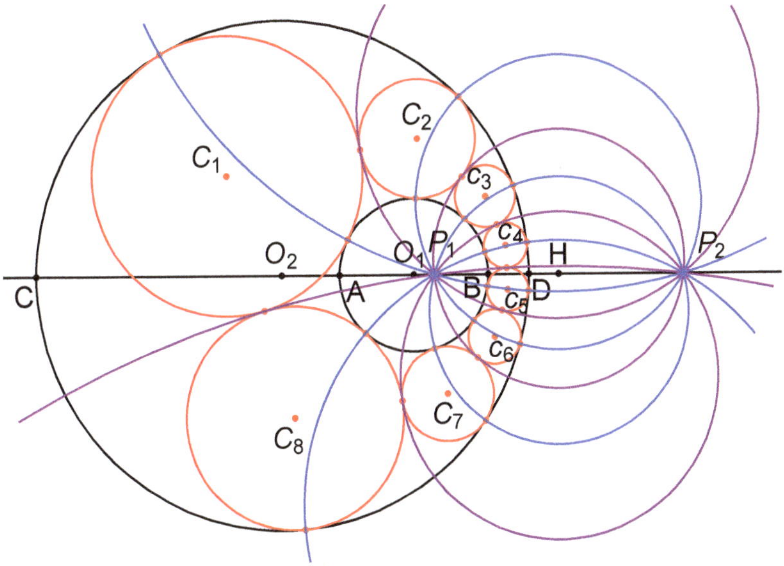

Figure 3.32: Steiner Chain with Base Circles (O_1) **and** (O_2), **for** $m = 1$ **and** $n = 8$

This follows from the fact that all these circles are orthogonal to both base circles (O_1, r) and (O_2, R), with $R > r$, and since these circles pass through one Poncelet point, they also pass from the other one (as it is known from the properties of the Poncelet points). This orthogonality follows from the fact that these circles are inverse images of straight lines that appear as diameters of the concentric base circles of the canonical chain that corresponds to the given Steiner chain, under an inversion with center one Poncelet point, see **Figure 3.33**. Also, we keep in mind the property of inversions that they preserve the absolute values of angles.

(2) Another property of the closed (finite, periodic) Steiner chains is depicted in **Figures 3.35** and **3.36**.

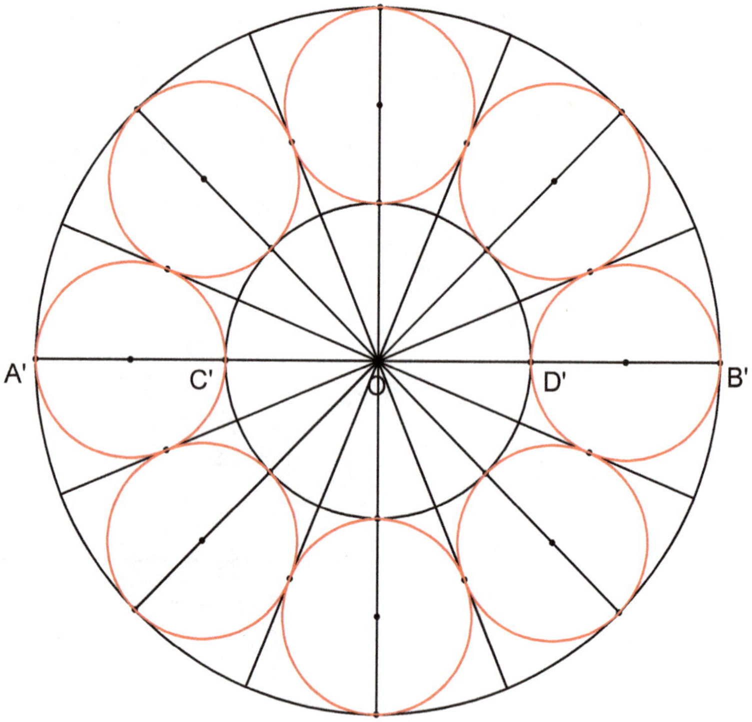

Figure 3.33: Canonical Steiner Chain for $m = 1$ **and** $n = 8$

In **Figure 3.35**, we have a Steiner chain with even number of circles ($n = 12$). We see that all common tangents of all pairs of successive chain circles at the common points of tangency are radial lines of the circle (L, q) and so they are concurrent at the center L of (L, q). [This fact holds for any Steiner chain because as we have seen the circle (L, q) always exists.]

We next see that the center lines of the pairs of opposite circles, all intersect at one point on the center line of the base circles.

In **Figure, 3.36** ($n = 12$ again), besides the center lines of the pairs of opposite circles, we have also drawn three additional groups of lines between the pairs of opposite circles and all these lines of the four groups intersect at one point, S, on the center line of the base circles.

Notice and prove that this common point of all these lines is the center of the positive homothety of the base circles (O_1, r) and (O_2, R), and so

$$\frac{SA}{SC} = \frac{SB}{SD} = \frac{r}{R},$$

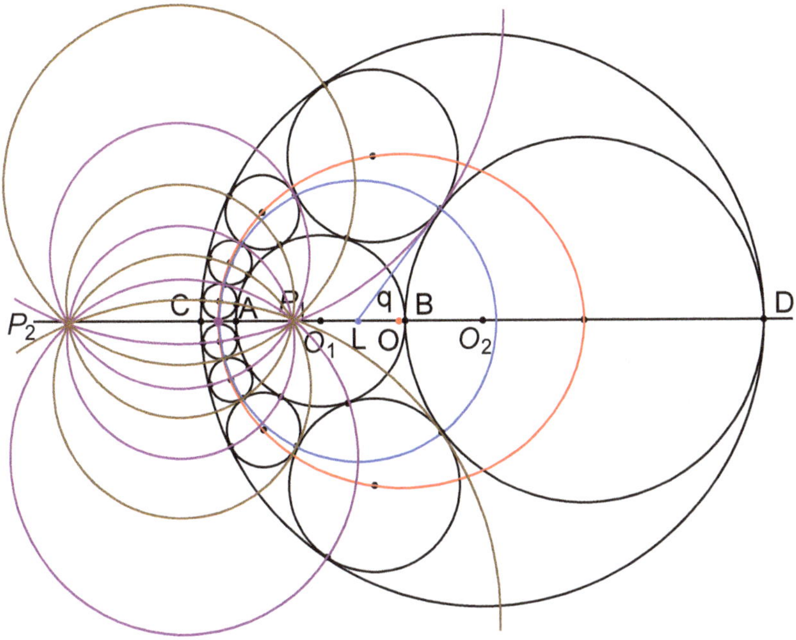

Figure 3.34: Steiner Chain with Base Circles (O_1) and (O_2), for $m = 1$ and $n = 9$

and so on. In this situation, S is also the center of the negative inversion of the circles (O_1, r) and (O_2, R) and also the homothetic center of the opposite homothety of the circles of each pair of opposite circles and in turn the center of their positive inversion. In these three new groups of lines, the lines join the corresponding points of tangency found on pairs of opposite circles. (Examine **Figure, 3.36**.)

We will justify this general result for Steiner chains with even number of circles. (This holds also when the base circles are outside of each other or if one circle is a straight line. We leave this to the reader as an **exercise**. Probably, someone can find a more elegant justification of this result.) We give our justification by using **Figures 3.37** and **3.38**. We try to avoid the metric relations based on the previously proven formulae, which are lengthy and involved. So, we will give a synthetic argument based on the properties of inversion and homotheties.

In **Figure 3.37**, we have a Steiner chain of eight circles, with nested base circles (O_1, r) and (O_2, R) (non-concentric) and with $r < R$. Along with this chain, we have put the corresponding canonical chain with center

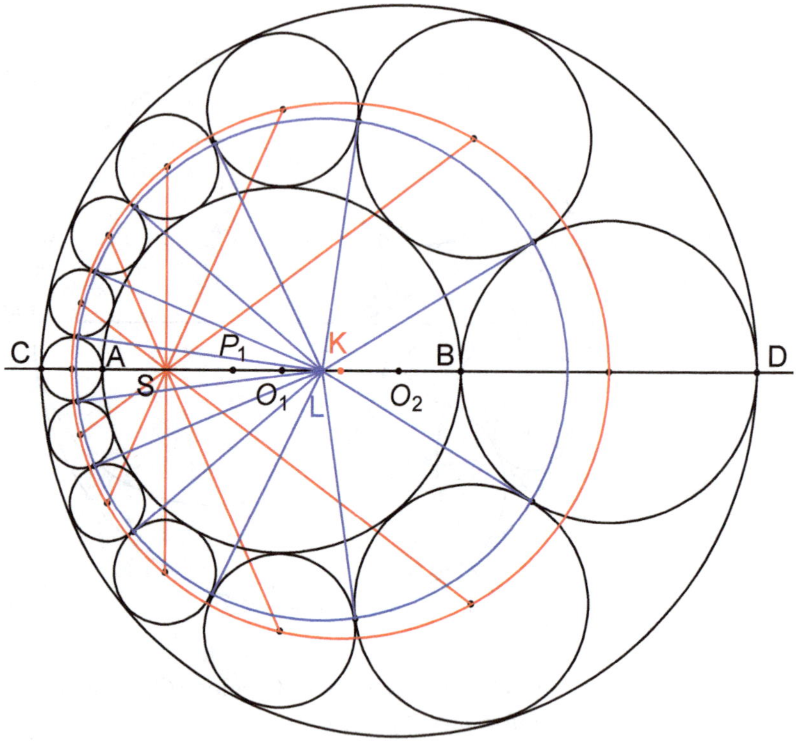

Figure 3.35: Steiner Chain with Base Circles (O_1) and (O_2), for $m = 1$ and $n = 12$

O', which is obtained by an inversion with center the Poncelet point P_1, which lies inside both base circles, and a power $c > 0$.

We also indicate how to find the homothetic centers of the positive homothety of the base circles (O_1, r) and (O_2, R) and the center of the negative homothety of the opposite circles of the chain (c_1) and (c_5). Then, we must prove that these two centers coincide at the same point on the center line O_1O_2, that we call it S.

(The point of intersection of the center line of two circles with the line defined by two parallel radii of the same direction is the center of the positive homothety. For the center of the negative homothety, we use parallel radii of opposite direction.)

In the next **Figure, 3.38**, we choose any two opposite circles of that Steiner Chain, e.g., (c_1) and (c_5), whose inverses are the circles of the

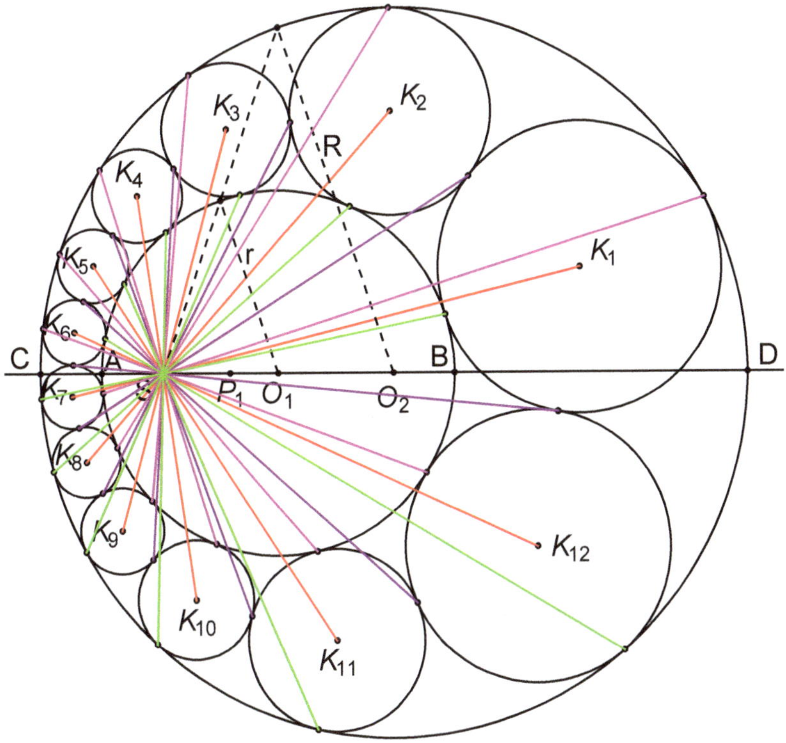

Figure 3.36: Steiner Chain with Base Circles (O_1) and (O_2), for $m = 1$ and $n = 12$

canonical chain (k_1) and (k_5), respectively. Let S be their homothetic center of their positive homothety.

The circles (c_1) and (k_1) are inverse of each other by the inversion with center P_1 and some positive power $c > 0$, and so, they are homothetic by a positive homothety with center P_1. The same fact is true for the circles (c_5) and (k_5). But, any opposite circles of the canonical chain, here the circles (k_1) and (k_5), are homothetic by the negative homothety with center O' and homothetic ratio -1. Composing these three homotheties in the correct order

$$(c_1) \longrightarrow (k_1) \longrightarrow (k_5) \longrightarrow (c_5),$$

we obtain a homothety mapping (c_1) to $c_5)$ and whose center is on the line P_1O', which is the same as the center line O_1O_2 of the two base circles. Therefore, the homothetic center S of (c_1) and (c_5) is the point of

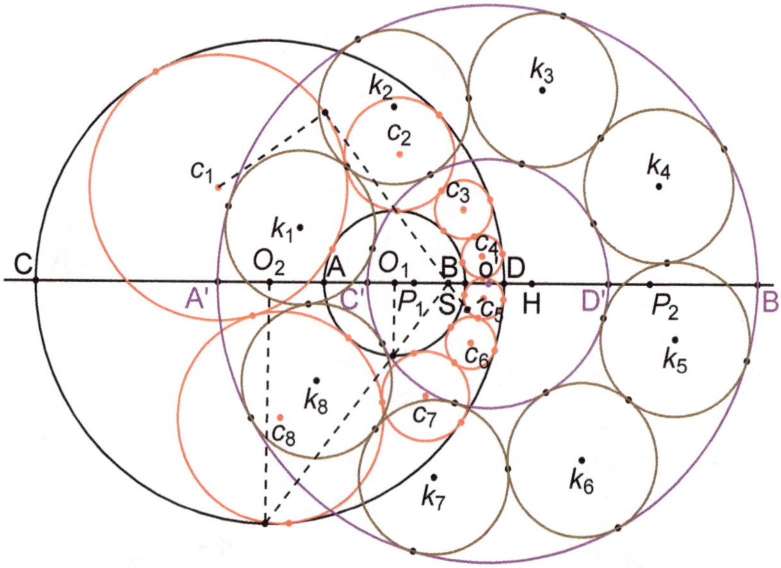

Figure 3.37: Steiner Chain with Base Circles (O_1) and (O_2), and its inverse canonical chain, with $m = 1$ and $n = 8$

intersection of the lines c_1c_5 and O_1O_2.

The center of the positive homothety of the base circles (O_1, r) and (O_2, R) is also located on the center line O_1O_2. If we prove that this is also the point S, then our original claim is proven.

Let us generically denote with $I_{[N,p]}$ the inversion with center N and power p. According to this notation, we have:

$$I_{[P_1,c]}(t_1) = t_1', \qquad I_{[O',-R_1'\cdot R_2']}(t_1') = s_5', \qquad I_{[P_1,c]}(s_5') = s_5, \qquad \text{that is,}$$

$$t_1 \quad \xrightarrow{I_{[P_1,c]}} \quad t_1' \quad \xrightarrow{I_{[O',-R_1'\cdot R_2']}} \quad s_5' \quad \xrightarrow{I_{[P_1,c]}} \quad s_5$$

By the properties of inversion, the composition

$$F = I_{[P_1,c]} \circ I_{[O',-R_1'R_2']} \circ I_{[P_1,c]}$$

is an involutory transformation (that is, $F \circ F = Id$, the identity; check this!) and the point

$$S = I_{[O',-R_1'R_2']} \circ I_{[P_1,c]}(P_1),$$

is mapped to $F(S) = \infty$ (check!).

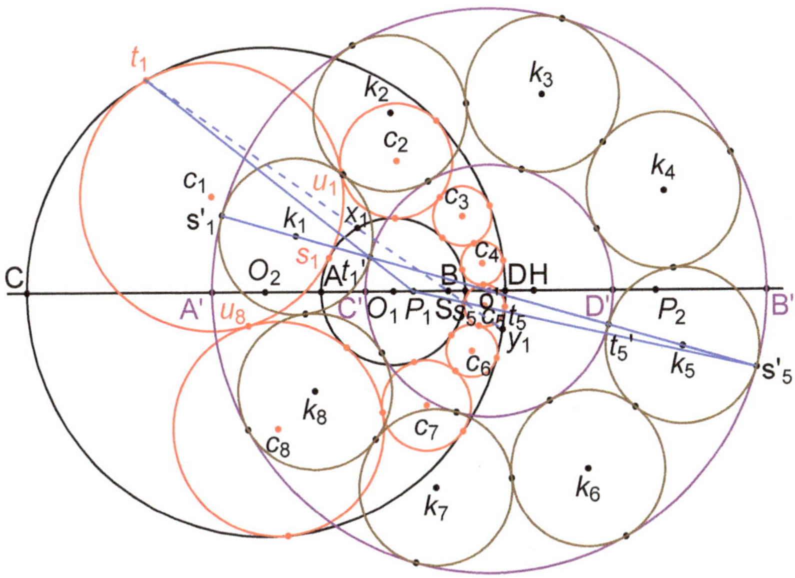

Figure 3.38: Steiner Chain with Base Circles (O_1) and (O_2), with $m = 1$ and $n = 8$. S is the homothetic center

{**Note:** To better understand what is going on, one may need to study first **Section 4.5**, in which we analyze the compositions of inversions of the type

$$I_{[O,R_1^2]} \circ I_{[Q,R_2^2]} \circ I_{[O,R_1^2]}.$$

That section may be considered as an appendix to the work we do here. This composition is a new inversion, whose circle is completely determined.}

Then, F is the new inversion $F = I_{[S,p]}$ with center the above point S, and power

$$p = \overline{SP_1} \cdot \overline{SQ}, \qquad \text{where} \qquad Q = I_{[P_1,c]}(O'),$$

and O' is the common center of the images of the two base circles under the inversion $I_{[P_1,c]}$, that maps them onto two concentric circles (O', R') and (O', r').

Now, we have $F(t_1) = s_5$. But t_1 is also on the circle (O_2, R) and point s_5 is on the circle (O_1, r). Also, the straight line $t_1 s_5$ contains another point x_1 of the circle (O_1, r) and point y_1 of the circle (O_2, R), and these two points are homothetic. (Any line containing two inverse points of two inverse circles, it contains only these points, when the line is tangent, and it contains two pairs of homologous homothetic points, otherwise, where the center of the homothety is the center of the inversion.) This holds the same for any pairs of points of tangency t_i and s_{i+4} in our figure. (In general, we have t_i and s_{i+k}, where $n = 2k$ and n is the number of circles in the closed chain.) Therefore, the point of intersection of the lines x_1y_1 and O_1O_2 is the center of homothety of the base circles (O_1, r) and (O_2, R) and this is the point S. This is the center of the positive homothety of (O_1, r) and (O_2, R).

Formulate and prove the analogous result when the closed (periodic) Steiner chain contains an odd number of circles ($m = 1$ and $n = 9$). In **Figure 3.39**, we have drawn two groups of lines that are concurrent at the point of the positive homothety of the base circles. Make a corresponding canonical chain for $m = 1$ and $n = 9$ and draw the related diameters, as was done in **Figure 3.33**. Can we find more groups of lines through that common point S, in this case?

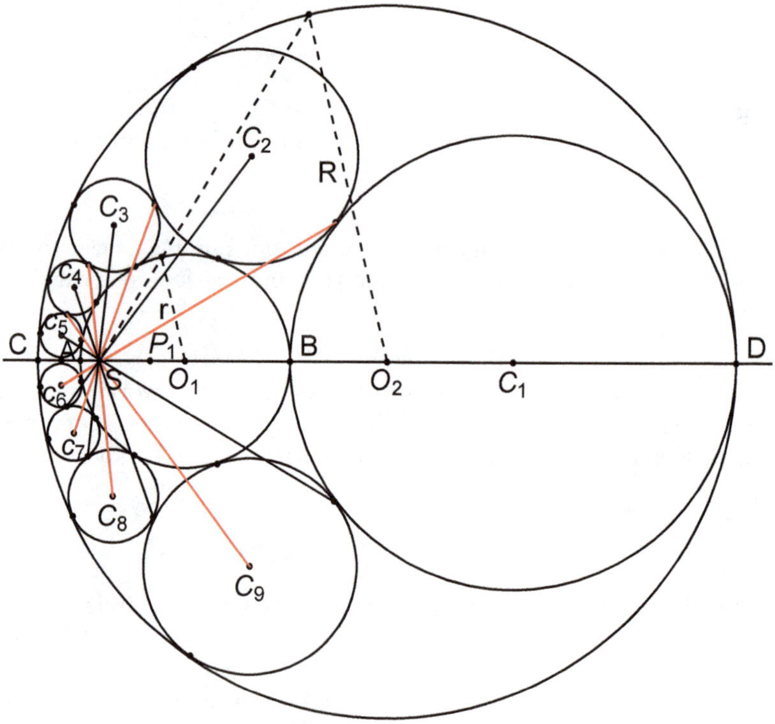

Figure 3.39: Steiner Chain with Base Circles (O_1) and (O_2), for $m = 1$ and $n = 9$

3.8 Important Related Geometrical Loci

It is an easy observation that the locus of the centers of all circles that are tangent to two intersecting straight lines is the union of the two bisectors of the angles of the intersecting lines, and the bisectors are perpendicular to each other. If the two straight lines are parallel, then this locus is their mid-parallel line. See **Figure 3.40**. Suppose now that we replace the two lines by two circles or by one line and one circle. Then, we want to find the analogous loci.

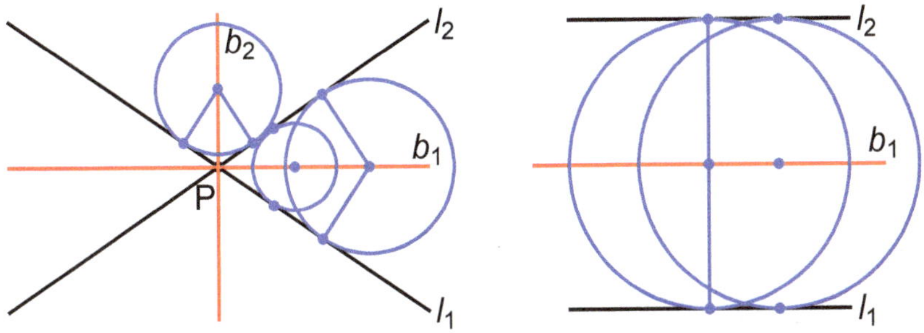

Figure 3.40: Locus of Equidistant Points from two Straight Lines

We have seen hyperbolas in **Figures 2.12, 2.18, 3.13, 3.16**, which were completely determined. Next, we are going to investigate the hyperbolas as certain loci with respect to two given circles such that no circle is completely inside the other one.

So, we firstly pose the following:
Problem: *Two circles* (O_1, R_1) *and* (O_2, R_2) *are given and such that no circle is completely inside the other. We want to find the locus of the centers of all circles that are tangent to the given circles* (O_1, R_1) *and* (O_2, R_2).

We will use the generic **Figure 3.41** in which the two circles, (O_1, R_1) and (O_2, R_2), are disjoint and exterior to each other. In this figure, we have $R_2 > R_1$. (Analogous results are obtained when the two circles are externally tangent or intersect at two points, as we have partially seen already in some of the figures stated above.) There are four cases for any circle which is tangent to two given such circles:
(1) externally tangent to both (O_1, R_1) and (O_2, R_2);
(2) internally tangent to both (O_1, R_1) and (O_2, R_2);
(3) externally tangent to (O_1, R_1) and internally tangent to (O_2, R_2);

(4) internally tangent to (O_1, R_1) and externally tangent to (O_2, R_2).

(1) Consider a circle externally tangent to both given circles with center C_1. As in **Figure 3.41**, we have

$$C_1X_1 = C_1X_2.$$

Then,

$$C_1O_2 - C_1O_1 = C_1X_1 - R_1 - (C_1X_2 - R_2) = R_2 - R_1 > 0$$

fixed. Therefore, C_1 belongs to the left branch of the hyperbola with foci O_1 and O_2 and major **semi-axis**

$$a = \frac{R_2 - R_1}{2}.$$

The **eccentricity** is

$$e = \frac{O_1O_2}{2a}.$$

Any point P of the left branch of this hyperbola has the same property as the center C_1 of the circle (C_1, C_1X_1). That is, there is a circle (P, r) which is externally tangent to both circles (O_1, R_1) and (O_2, R_2). (We have depicted this hyperbola as a solid curve.)

(2) Consider a circle internally tangent to both given circles with center C_2. Then, $C_2Y_1 = C_2Y_2$ and

$$C_2O_1 - C_2O_2 = (C_2Y_1 - R_1) - (C_2Y_2 - R_2) = R_2 - R_1.$$

Therefore, C_2 belongs to the same hyperbola found above, but cannot belong to the left branch. Thus C_2 belongs to the right branch of this hyperbola.

Also, any point Q of the right branch of this hyperbola has the same property as the point C_2. That is, there is a circle (Q, ρ) which is internally tangent to both circles (O_1, R_1) and (O_2, R_2).

Cases (3) and (4) yields one more hyperbola, the dashed one in the **Figure 3.41**. This hyperbola has the following property: the circles which are internally tangent to one of the given circles and externally to the other have centers on this hyperbola. The foci are again O_1 and O_2, but the major

$$a = \frac{R_2 + R_1}{2}.$$

Indeed, if for the point C_1', we have $C_1'X_1' = C_1'X_2'$, then

$$C_1'O_2 - C_1'O_1' = C_1'X_2' + R_2 - (C_1X_1' - R_1) = R_2 + R_1,$$

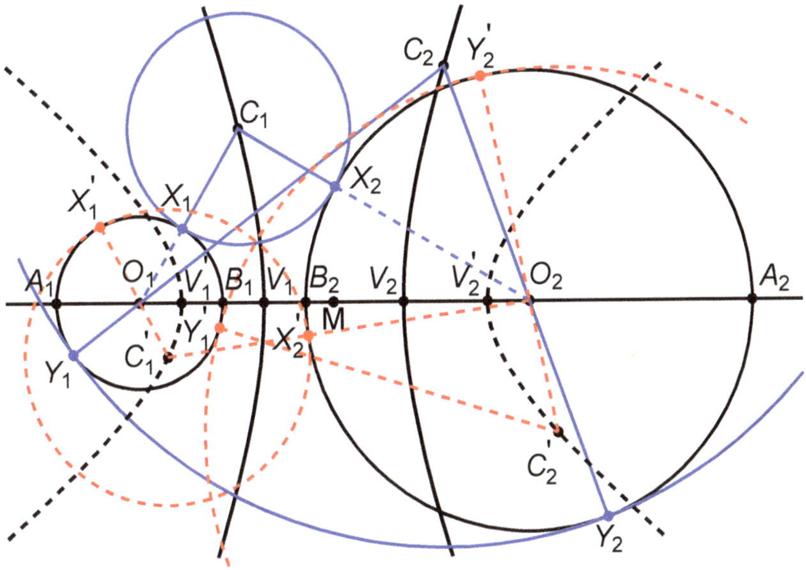

Figure 3.41: Two Hyperbolas as Locus Between Two Circles.

and similarly for the point C_2' in the figure.

Finally, the locus of the problem stated above is the set of these two hyperbolas.

We have also seen ellipses as certain loci with respect to two given circles such that one is inside the other. See, **Figures 2.1, 2.2, 2.3, 2.6, 2.7, 2.18, 3.4, 3.5, 3.6**. Even though these ellipses were completely determined, we will see some additional information about them, in some of the following remarks.

Remarks. (1) The vertex V_1 is the midpoint of B_1B_2 and the vertex V_2 is the midpoint of A_1A_2. Similarly, the vertex V_1' is the midpoint of A_1B_2 and the vertex V_2' is the midpoint of B_1A_2. (Prove!)

(2) If $R_1 = R_2$, the solid hyperbola degenerates to the midpoint perpendicular straight line of the segment O_1O_2, since in this case, we would have $CO_1 = CO_2$.

(3) If (O_1, R_1) and (O_2, R_2) are externally tangent, then the point of tangency is the vertex of one of the branches of the solid hyperbola. See **Figure 2.12**.

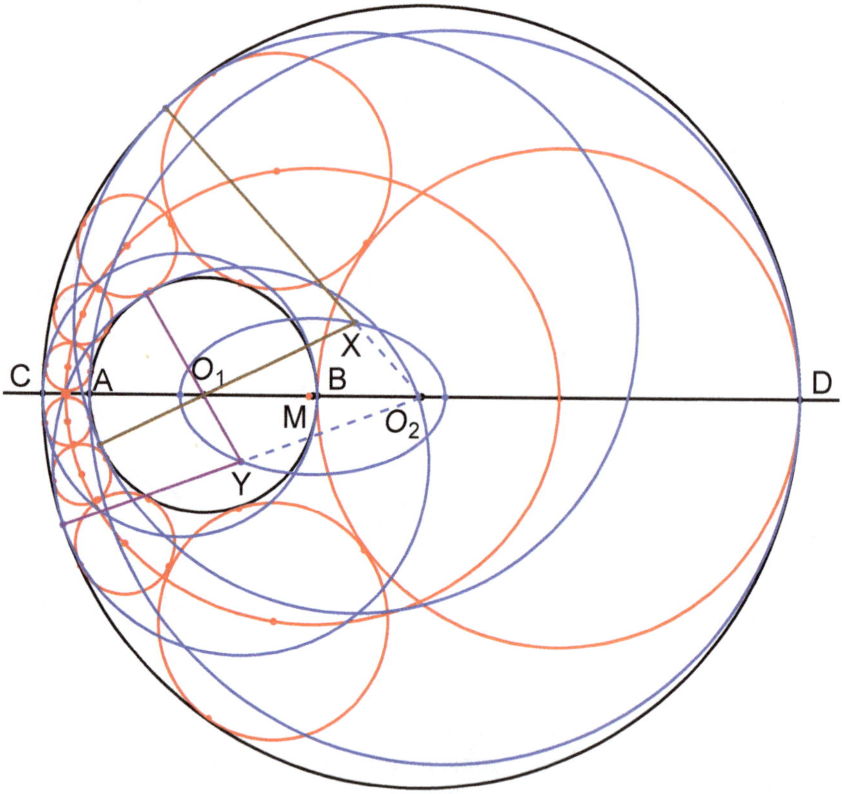

Figure 3.42: Two Ellipses as Locus Between Two Nested Circles.

(4) If (O_1, R_1) and (O_2, R_2) intersect at two points, then one of the branches of the solid hyperbola passes through both of them and the other hyperbola changes to an ellipse. See **Figure 2.18** and determine both the hyperbola and the ellipse.

(5) If (O_1, R_1) lies inside (O_2, R_2), then the two hyperbolas change to two ellipses, with major semi-axes

$$a = \frac{R_2 + R_1}{2} \quad \text{and} \quad a = \frac{|R_2 - R_1|}{2}$$

and the same foci O_1 and O_2 for both. See **Figure 3.42**.

(6) If (O_1, R_1) is internally tangent to (O_2, R_2), then one ellipse is drawn

in **Figures 2.2** and **2.6**. The smaller ellipse degenerates to the segment O_1O_2 and the circles that have centers the points of this segment must pass through the point of tangency of (O_1, R_1) and (O_2, R_2).

(7) If (O_1, R_1) and (O_2, R_2) are concentric (i.e., $O_1 = O_2 := O$), then the two ellipses become concentric circles with center O and radii $\frac{R_2 + R_1}{2}$ and $\frac{|R_2 - R_1|}{2}$. (Make a figure!)

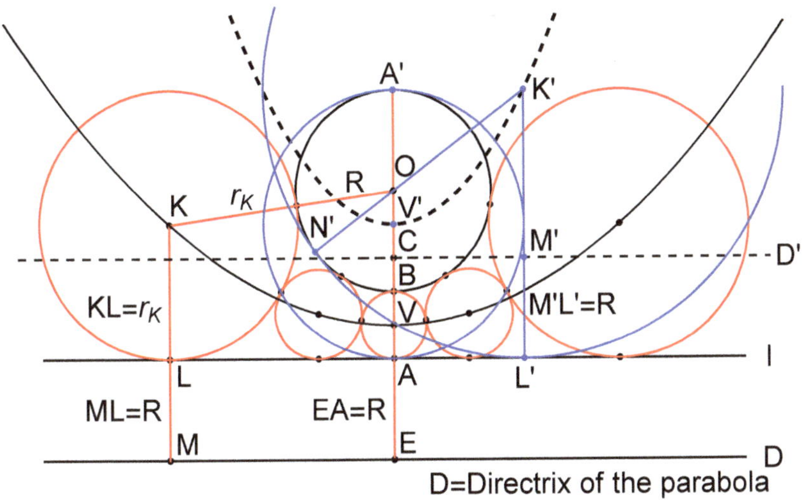

Figure 3.43: Two Parabolas as Locus for Circle and Straight Line.

Next, we investigate the following:
Problem: *A circle (O, R) and a straight line l are given. We want to find the locus of the centers of all circles that are tangent to the circle and straight line.*

We are going to see that the locus of all sought points is the set of two parabolas. Parabolas have occurred in such situations in **Figures 2.16, 2.23, 2.24, 3.9, 3.12** and **3.14**. The foci and directrices of these parabolas were found, and so, the parabolas were completely determined. Here, we generically examine **Figure 3.43**.

Suppose K is a point such that a circle (K, r_K) is externally tangent to the circle (O, R) and tangent to the straight line l. We then draw the line D parallel to l at distance R and on the half plane that does not contain the circle (O, R). Then,

$$KO = R + r_K = ML + LK = KM.$$

Therefore, the point K is a point of the parabola with focus O and directrix the line D. This is the solid parabola in **Figure 3.43** and every point of it is the center of a circle externally tangent to (O, R) and tangent to l.

Now we consider the line D' the reflection D in l and the parabola with focus O and directrix D' (the dashed parabola in the figure). Every point of this parabola is the center of a circle internally tangent to (O, R) and tangent to l. To see this, we draw the segment $K'ON'$ and the segment $K'M'L'$ perpendicular to and D'. Then, we have $K'O = K'M'$, and so

$$K'N' = K'O + R = K'M' + R = K'L'.$$

Remarks: (1) The vertex V of the solid parabola is the midpoint of OE and AB. The vertex V' of the dashed parabola is the midpoint of OC and AA'.

(2) If the line l is tangent to (O, R), as in **Figure 2.16**, then the solid parabola has vertex V the point of tangency, and the dashed parabola degenerates to the straight line OV, which is perpendicular to l. For any point X of the half line $[VO\infty)$, the circle (X, XV) is tangent to l and internally tangent to (O, R) at V. For any point X' of the half line $(-\infty V]$, the circle $(X', X'V)$ is tangent to l and externally tangent to (O, R) at V.

(3) The line l can intersect the circle (O, R) as in **Figures 2.23** and **2.24**. Study these figures and draw the corresponding parabolas.

(4) Along these lines we can ask for the locus of the centers of the circles that pass through a given point and are tangent to a given circle or a given straight line. This is a special case of the previous cases if one of the given circles has radius zero and so it becomes the given point. Depending on the position of the point with respect to the given circle or the straight line, the locus may be hyperbola, ellipse, circle, parabola, straight line.

In conclusion: *We see the connection of conic sections as geometrical loci with either the Pappus chains and / or the Steiner chains. We also see that in studying all variations of these loci all conic sections appear.*

3.9 Some Properties of the Circle $\mathcal{C}[L,q]$

In the theory of inversion, the following results are proven: Given two circles or one circle and a straight line then:

(a) If they intersect, then an inversion with pole one of the common points transforms them into two straight lines trough the image of the other common point.

(b) If they are tangent, then an inversion with pole the point of tangency transforms them into two straight parallel lines.

(c) If the two circles do not intersect, then an inversion with pole one of the Poncelet points transforms them into two concentric circles trough the image of the other Poncelet point.

(d) If the circle and the straight line do not intersect, then there is inversion that transforms them into two concentric circles. (See bibliography, e.g., Frère, G.-M. 1926, Pedoe, D. 1970, Roussos, I. M. 2022, Yaglom, I. M. 2009, etc. See also **Chapter 4.**)

In Roussos, I. M. 2022, Theorem 6.4.6, pp. 333-334, (see also **Chapter 4, Theorem 4.4.2**), we find pairs of inversions that have the same center but different powers that map two given circles into two equal circles. Those inversions have centers on the center line of the circles and appropriate powers. Here, we examine other inversions that mapped two given circles or a circle and a straight line into two equal circles by using the circle $\mathcal{C}[L,q]$ that contains the points of tangency of a chain of circles (Pappus or Steiner, open or closed, regardless) with base circles the two given circles or the given circle and the straight line. We obtain the following **property of** $\mathcal{C}[L,q]$**:** *An inversion of any power and with pole any point of this circle transforms the two base circles or the given circle and the straight line (either they intersect or do not) into two circles of equal radii (equal circles).*
See **Figures 3.44, 3.45, 3.46, 3.47**, and **3.48**. In these figures the base circles and the circles of the chain are black, the circle that contains the points of tangency is blue and the images equal circles are red.

We have seen the circle $\mathcal{C}[L,q]$, which contains the points of tangency of the circles of a chain, in all the chains that have been presented in the previous text. We have dealt with finding its center and radius a great deal. We also know that this circle is orthogonal to the circles of the chain. (See the figures that contain a chain of circles in the previous sections of this text.) If the base circles or circle with a line intersect, then this circle obviously passes through the common points and, as we will prove below the following **property**: ***The circle*** $\mathcal{C}[L,q]$ *forms equal angles with two tangent or intersecting base circles, or one base circle and a straight line.*
[See the figures that follow and **Section 2.5, Case (1)**.]

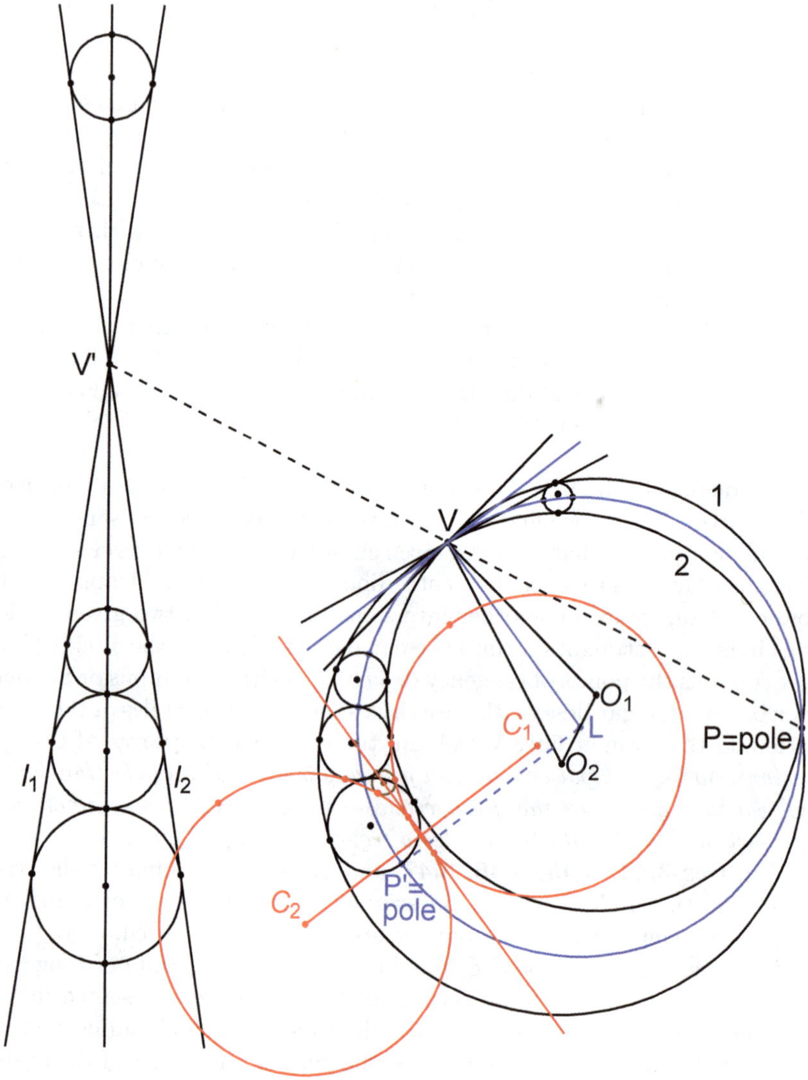

Figure 3.44: Inversion that Maps Two Intersecting Circles into Equal Circles.

Let us study now **Figure 3.44**. This study will extend analogously to the cases presented in the other figures. We have two circles (O_1) and (O_2) intersecting at the points V and P. Four circles of a chain with base circles (O_1) and (O_2) are also depicted along with the blue circle (L) that contains the points of tangency of the chain. An inversion with pole P transform (O_1) and (O_2) to straight lines l_1 and l_2 which intersect at point V', which is the inverse of V. The circles of the chain transform to circles successively tangent to each other and tangent to the lines l_1 and l_2. The circle L, being orthogonal to the circles of the chain, transforms to straight line through V' and contains the centers of the image circles. Therefore, this line is the bisector of the angle of the lines l_1 and l_2. Since inversions preserve the absolute values of oriented angles, this proves the fact that $\mathcal{C}[L, q]$ forms equal angles with the circles O_1 and O_2.

Now, we pick a point P' on the circle (L) and we perform an inversion of the circles O_1 and O_2 with pole P'. Then, the points P and V are mapped to the common points of the image circles (in red) and the circle (L) to their common chord. Since inversions preserve the absolute values of angles, the common cord bisects the angle of the intersecting image circles. This implies that the image circles are equal (fairly easy exercise to prove).

We notice another **property**:

The center line C_1C_2, of the two equal image circles, is parallel to the line $P'L$ (joining the chosen pole-center of the inversion on $\mathcal{C}[L, q]$ to the center L of the $\mathcal{C}[L, q]$).

This is so because:

(1st) the circle $\mathcal{C}[L, q]$ is inverted to the common chord of the two equal image circles (C_1) and (C_2), and so it is perpendicular to C_1C_2;

(2nd) the radius LP' is perpendicular to the circle $\mathcal{C}[L, q]$, and it is inverted onto itself and so it is perpendicular to the common chord (or the common tangent, as we see in **Figure 3.46**, and any other adjustment in other figures that follow).

Next, we study **Figure 3.45**. We have two circles (O_1) and (O_2), with (O_1) inside (O_2) and with these circles as base circles, we have drawn three circles of a chain (Steiner chain). The circle that contains the points of tangency of the chain circle is $\mathcal{C}[L, q]$. We pick a point P on $\mathcal{C}[L, q]$ as the pole of an inversion. The inverses of the circles (O_1) and (O_2) are the circles $\mathcal{C}(C_1, R_1)$ and $\mathcal{C}(C_2, R_2)$. We have also drawn the dashed circles, the inverses of the three circles of the chain. The inverse of $\mathcal{C}[L, q]$ is the straight line that contains the centers $\{K_1,\ K_2,\ \ldots\}$ and the tangent points of the dashed circles. Since the center L is on O_1O_2, this straight line is perpendicular to C_1C_2 at a point M.

We will prove that $R_1 = R_2$ and M is the midpoint of the center segment C_1C_2. The argument for the previous **Figure, 3.44,** cannot apply now to

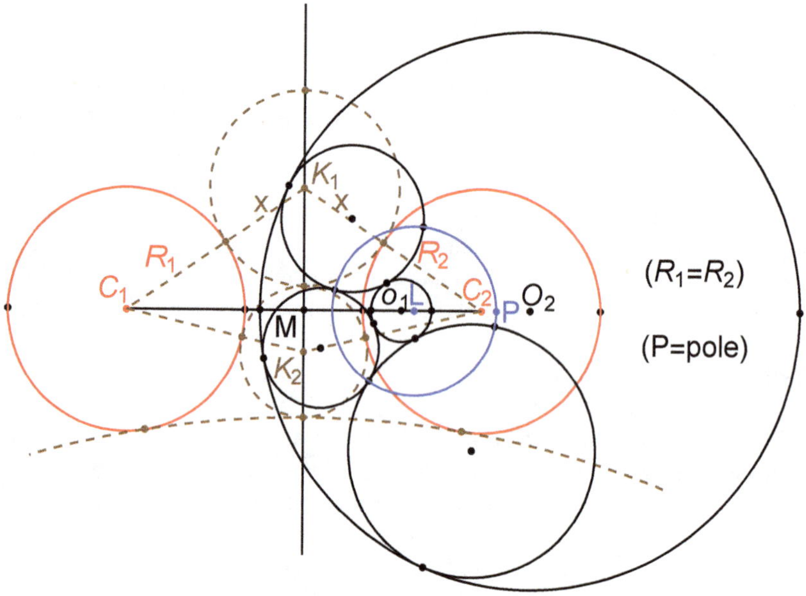

Figure 3.45: Inversion that Maps Two non-Intersecting Circles into Equal Circles.

Figure 3.45, as $\mathcal{C}(C_1, R_1)$ and $\mathcal{C}(C_2, R_2)$ do not intersect. Thus, we work as follows:

Let x be the radius of any dashed circles, here for the sake of argument of (K_1). We let

$$a = C_1M, \qquad b = MC_2 \quad \text{and} \quad h = MK_1.$$

Since all the centers of the dashed circles are on a straight line perpendicular to the center line O_1O_2, we have:

$$(R_1 + x)^2 = a^2 + h^2 \qquad \text{and} \qquad (R_2 + x)^2 = b^2 + h^2.$$

We subtract these two equations and we get the linear equation

$$2(R_1 - R_2)x + R_1^2 - R_2^2 + a^2 - b^2 = 0.$$

Since in a chain there are at least two circles tangent to each other, this linear polynomial has at least two solutions, and so it must be identically zero. Hence,

$R_1 = R_2$, and then, $a = b$, and so, M is the midpoint of C_1C_2.

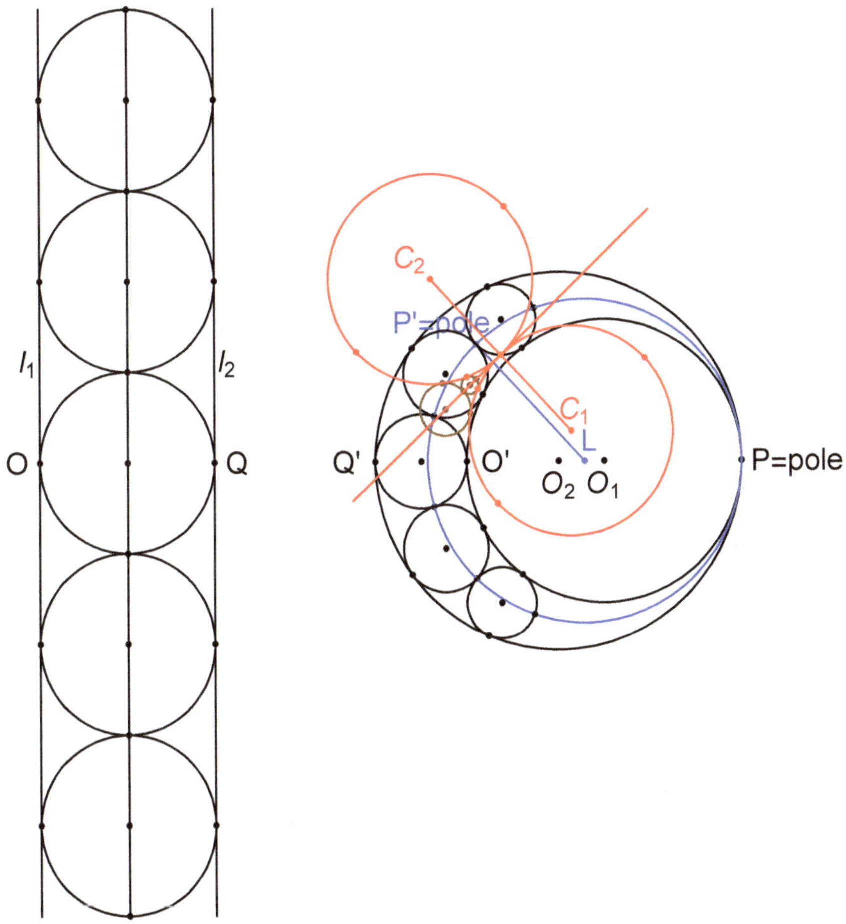

Figure 3.46: Inversion that Maps Two internally Tangent Circles into Equal Circles.

This argument also works in the three situations of **Figures, 3.47, 3.48**, and **3.49**, at the end of this section, in which we have:

(1) a Pappus chain between two internally tangent circles,

(2) a Pappus chain between a circle and one of its tangent lines, and

(3) a closed Steiner chain of six circles with the two base circles exterior to each other.

(We hope that we have given enough information and the figures are self-explanatory for any reader who knows classical geometry and inversion to understand what is going on in the next three figures and also make other figures that, for the sake of brevity, are not included here, and with the other kinds of chains we have provided in previous sections of the text.)

Remarks: (1) Since inversions are involutary (i.e., inverse of themselves, and so, $I \circ I = Id$), the figures provided show a way how to construct chains analogous to the chains that are depicted in the figures. (Explain this to yourselves.)

(2) The above ideas work even when we have two straight lines and we want to invert them into two equal circles. (Write out the details.)

(3) In the previous sections, we have shown how we figure out the center and the radius of an image circle under an inversion. We have also determined the circle(s) $\mathcal{C}(L, q)$('s). So, choosing a point P' or P on this circle as a pole (center) of the inversion and some power c, we can easily determine the centers O_1 and O_2 and the radius of the equal image circles. (Work this out!)

Four Suggested Problems:
(a) Find the loci of the points C_1 and C_2 as the pole of the inversion P moves along the circle $\mathcal{C}(L, q)$ and the power of the inversion is fixed.
(b) Find the loci of the points C_1 and C_2 as the pole of the inversion P is fixed and the power of the inversion varies in some concrete continuous way.
(c) Find the loci of the points C_1 and C_2 as the pole of the inversion P moves along the circle $\mathcal{C}(L, q)$ and the power of the inversion varies in some concrete continuous way.
(d) Can any point of the plane be C_1, (or C_2) for some pole and some power?

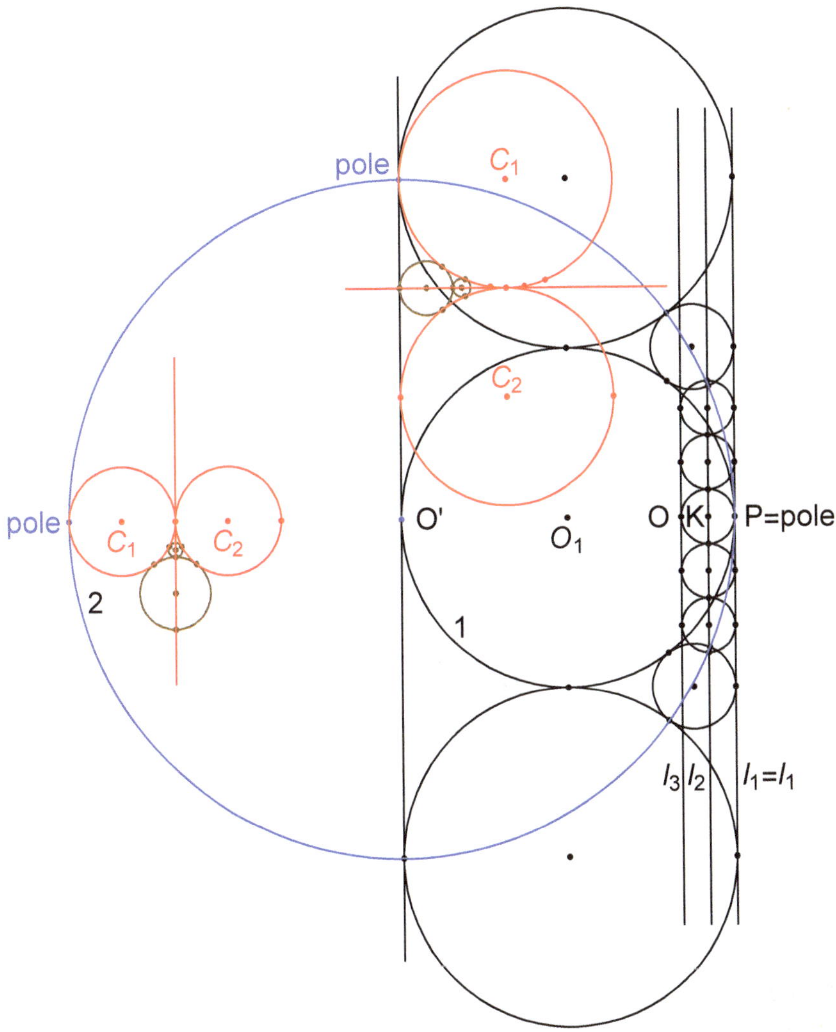

Figure 3.47: Two Inversions that Maps a Circle and a Tangent Line into Equal Circles.

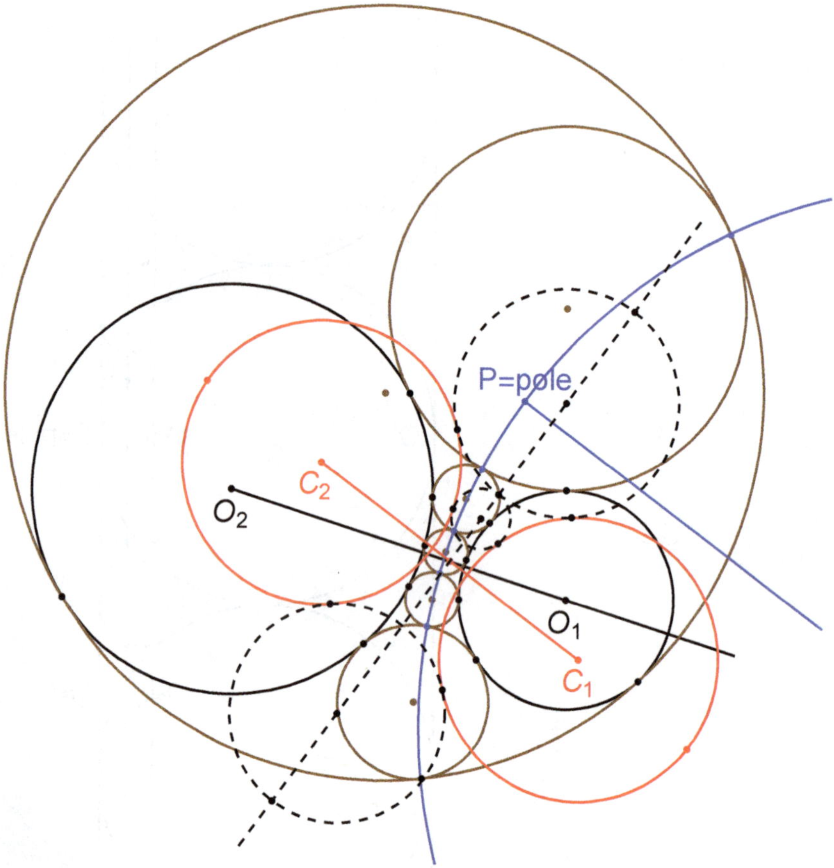

Figure 3.48: Inversion that Maps two Circles Outside One Another into Equal Circles.

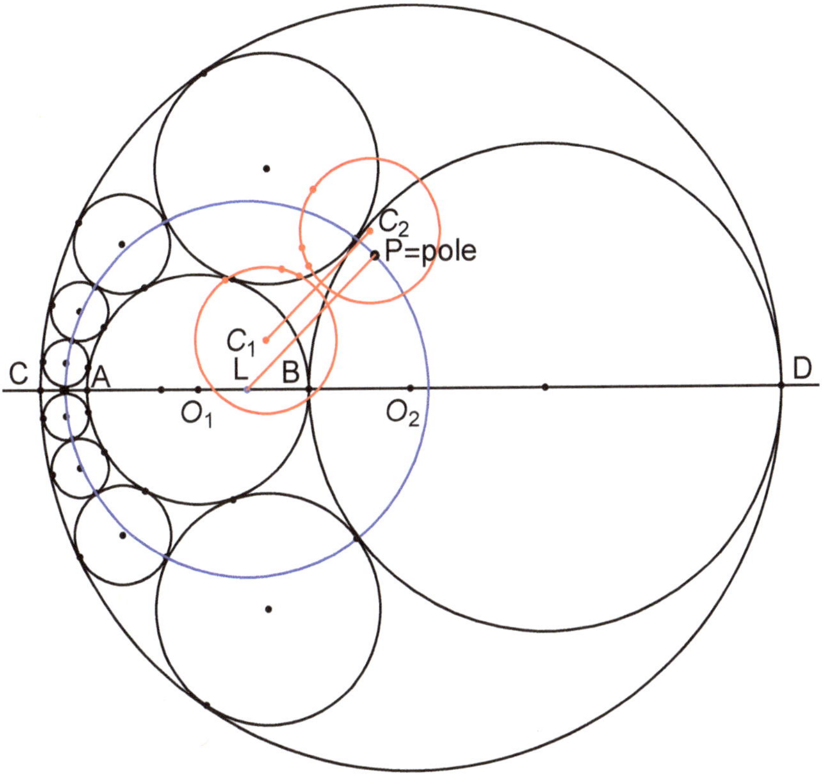

Figure 3.49: Inversion that Maps two Circles Inside One Another into Equal Circles.

Chapter 4

Appendix of Prerequisites

To help the unfamiliar reader, in this appendix, we provide the most important theorems and results that we have referred to and use in the previous material. We arrange them in three groups, each of which contains related topics.

4.1 Three Theorems

(1) We begin with the **theorem of medians** in a triangle following theorem of medians which is used in understanding the radial axis of two circles.

Theorem 4.1.1 (Theorem of Medians) *We consider any triangle ABC its median $m_a = AD$ (D is the midpoint of BC), ED the projection of this median on the side $a = BC$ (E is the foot of the altitude on $a = BC$ from the vertex A) and, without loss of generality, we may assume that $b = AC \geq c = AB$. Then we have:*

$$(a) \quad b^2 + c^2 = 2 \cdot m_a^2 + \frac{a^2}{2},$$

$$(b) \quad b^2 - c^2 = 2 \cdot a \cdot ED,$$

$$(c) \quad a^2 = 4[m_a^2 - bc\cos(\widehat{A})].$$

Analogous results for the other medians are obtained by a cyclic permutation of the letter a, b, and c.

Proof See **Figure 4.1**.

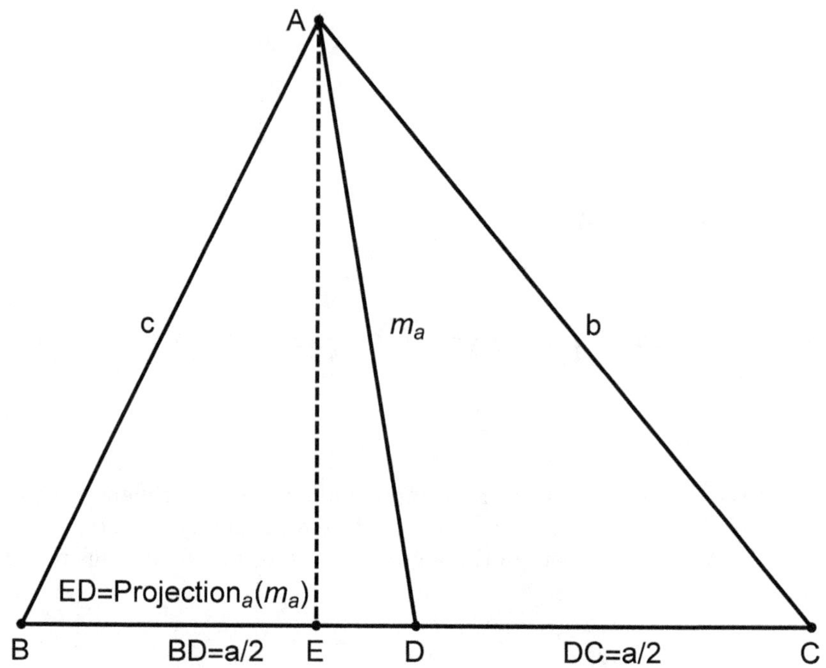

Figure 4.1: Theorem of Medians

If $b = c$, i.e., the triangle ABC is isosceles, then $E = D$, i.e., $ED = 0$ and height $AE = AD = m_a$ median. Then, the relations **(a)** and **(b)** follow immediately form the Pythagorean Theorem applied to the right triangles ACE and ABE.

If $b > c$, then E and B are on the same side of D along the line BC. We use the Pythagorean Theorem again and from the right triangles ACE and ABE and the relation $BD = CD = \frac{a}{2}$, we get

$$b^2 = AE^2 + CE^2 = AE^2 + \left(\frac{a}{2} + ED\right)^2 = AE^2 + \frac{a^2}{4} + a \cdot ED + ED^2,$$

$$c^2 = AE^2 + BE^2 = AE^2 + \left(\frac{a}{2} - ED\right)^2 = AE^2 + \frac{a^2}{4} - a \cdot ED + ED^2.$$

By adding and subtracting the two equalities sidewise, and using the right triangle AED ($\widehat{E} = 90°$, $AE^2 + ED^2 = AD^2 = m_a^2$), we find the two results

$$\textbf{(a)} \quad b^2 + c^2 = 2\left(AE^2 + ED^2\right) + \frac{a^2}{2} = 2 \cdot m_a^2 + \frac{a^2}{2},$$

$$\textbf{(b)} \quad b^2 - c^2 = 2 \cdot a \cdot ED.$$

Result **(c)** follows from **(a)** and the law of cosines.

■

Corollary 4.1.1 *In any triangle ABC we have*

$$(a) \quad m_a^2 = \frac{2(b^2 + c^2) - a^2}{4} \qquad \Longleftrightarrow \qquad m_a = \frac{\sqrt{2(b^2 + c^2) - a^2}}{2},$$

$$(b) \qquad b^2 - c^2 = 2a \cdot \overline{ED} \qquad \Longleftrightarrow \qquad \overline{ED} = \frac{b^2 - c^2}{2a},$$

$$(c) \qquad \overline{BE} = \frac{a}{2} - \overline{ED} = \frac{a}{2} - \frac{b^2 - c^2}{2a} = \frac{a^2 - b^2 + c^2}{2a},$$

$$(d) \qquad \overline{EC} = \frac{a}{2} + \overline{ED} = \frac{a}{2} + \frac{b^2 - c^2}{2a} = \frac{a^2 + b^2 - c^2}{2a}.$$

Analogous results for the other medians are obtained by a cyclic permutation of the letter a, b, and c.

(2) Now we can easily prove **Heron's Formula.**

Using **(c)**, after simplifying we get

$$AE^2 = c^2 - BE^2 = c^2 - \left(\frac{a^2 - b^2 + c^2}{2a}\right)^2.$$

Setting

$$s = \frac{a + b + c}{2}$$

(half of the perimeter of ABC), and expanding we find that AE is

$$h_a = \frac{2}{a}\sqrt{s(s - a)(s - b)(s - c)}. \tag{4.1}$$

This is **Heron's formula** for the height $h_a := AE$. Therefore, the area of ABC is given by **Heron's formula** formula for the area

$$\text{Area}(ABC) = \sqrt{s(s-a)(s-b)(s-c)}.$$

(3) Generalization of the Theorem of Medians is **Stewart's Theorem**. We have used this Theorem earlier, and so, we will state, prove, and generalize it below. (We can find the lengths of medians, heights, angle bisectors, and so on, of a triangle, by using this Theorem.)

Theorem 4.1.2 (Stewart's Theorem) *We consider any triangle ABC and a line AD intersecting the side BC at a point D between E and C. Then Stewart's relation holds,*

$$AB^2 \cdot DC + AC^2 \cdot BD = BC \cdot AD^2 + BC \cdot BD \cdot DC.$$

Proof See **Figure 4.2**. If we use the law of cosines we find,

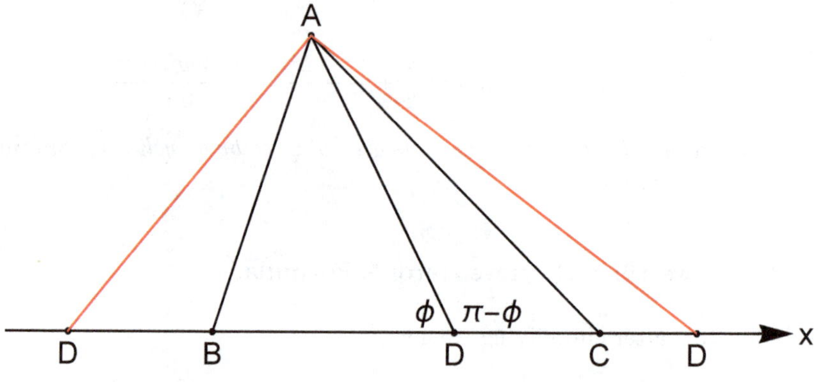

Figure 4.2: For Stewart's Theorem

$$AB^2 = AD^2 + BD^2 - 2AD \cdot BD \cos(\phi)$$

$$AC^2 = AD^2 + DC^2 - 2AD \cdot DC \cos(\pi - \phi) = AD^2 + DC^2 + 2AD \cdot DC \cos(\phi),$$

from which we get

$$AB^2 \cdot DC = AD^2 \cdot DC + BD^2 \cdot DC - 2AD \cdot BD \cdot CD\cos(\phi)$$

$$AC^2 \cdot BD = AD^2 \cdot BD + DC^2 \cdot BD + 2AD \cdot DC \cdot BD\cos(\phi).$$

We add the last two relations sidewise and we find

$$AB^2 \cdot DC + AC^2 \cdot BD = AD^2 \cdot DC + BD^2 \cdot DC + AD^2 \cdot BD + DC^2 \cdot BD =$$
$$(BD + DC)AD^2 + (BD + DC)BD \cdot DC = BC \cdot AD^2 + BC \cdot BD \cdot DC,$$

that is Stewart's relation

$$\text{(a)} \quad AB^2 \cdot DC + AC^2 \cdot BD = BC \cdot AD^2 + BC \cdot BD \cdot DC.$$

■

Now: If the point D lies on the left of B, in the same way, we obtain

$$\text{(b)} \quad AB^2 \cdot DC - AC^2 \cdot BD = BC \cdot AD^2 - BC \cdot BD \cdot DC,$$

and if D lies on the right of C, in the same way, we obtain

$$(c) \quad - AB^2 \cdot DC + AC^2 \cdot BD = BC \cdot AD^2 - BC \cdot BD \cdot DC.$$

Therefore, we have the following **generalization of Stewart's relation:**
We consider the straight line BC as an **axis** $\overrightarrow{x}$, (with positive direction from left to right, let us say) in the plane and we pick three points on it B, C, and D (in any order), and we let A be **any point** of the plane (even on the axis itself), then, using oriented segments all three above relations are written in one, in the following way:

$$AB^2 \cdot \overline{DC} + AC^2 \cdot \overline{BD} = \overline{BC} \cdot (AD^2 + \overline{BD} \cdot \overline{DC})$$

or

$$AB^2 \cdot \overline{DC} + AD^2 \cdot \overline{CB} + AC^2 \cdot \overline{BD} + \overline{BD} \cdot \overline{DC} \cdot \overline{CB} = 0.$$

(If all four points are on the axis and we introduce coordinates such that A be at a, B be at b, C be at c, and D be at d, and then replace the oriented segments, $\overline{AB}$ by $b - a$, etc., in the above general relation, we perform the computation and we shall see that this relation is valid, proving it analytically.)

4.2 Homotheties

There are two kinds of a **homothety** (or **dilation** or **dilatation**), the positive and the negative, according to the following definitions:

Definition 4.2.1 *(a) Given a point in the plane $O \in \mathcal{P}$ and a positive real number $\lambda > 0$, we call* ***positive plane homothety*** *with* ***center*** *O and* ***positive ratio*** *$\lambda > 0$, the plane mapping*

$$\mathcal{H}_{[O,\lambda]} \ : \ \mathcal{P} \longrightarrow \mathcal{P}$$

defined as follows: $\forall\ A \in \mathcal{P}$, we define $\mathcal{H}_{[O,\lambda]}(A) = A' \in \mathcal{P}$, such that, the point A' is on the ray OA (hence, O, A and A' are collinear and the points A and A' are on the same side of O) and

$$d[O, A'] = \lambda \cdot d(O, A), \qquad or \qquad OA' = \lambda \cdot OA.$$

See ***Figure 4.3.***

Figure 4.3: Positive homothety $\mathcal{H}_{[O,\lambda]}$, with ratio $\lambda = \dfrac{\overline{OA'}}{\overline{OA}}$

(b) Given a point $O \in \mathcal{P}$ and a negative real number $\mu < 0$, we call $\mathcal{H}_{[O,\mu]}$ ***negative plane homothety*** *with* ***center*** *O and* ***negative ratio*** *$\mu < 0$ if $\forall\ A \in \mathcal{P}$, we define $\mathcal{H}_{[O,\mu]}(A) = A' \in \mathcal{P}$, such that, the point A' lies on the ray opposite to OA (and so O, A and A' are collinear with O between A and A') and*

$$d[O, A'] = |\mu| \cdot d(O, A), \qquad or \qquad OA' = |\mu| \cdot OA.$$

See ***Figure 4.4.***

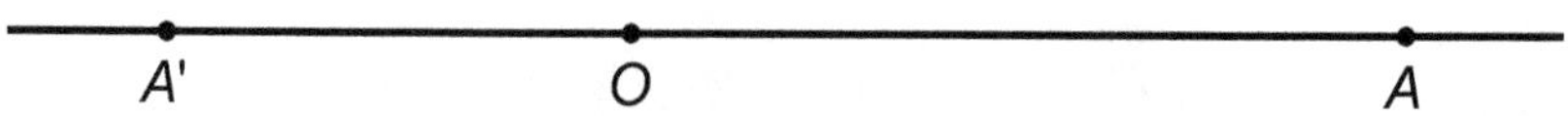

Figure 4.4: Negative homothety $\mathcal{H}_{[O,\mu]}$, with ratio $\mu = \dfrac{\overline{OA'}}{\overline{OA}}$

(c) For a homothety $\mathcal{H}_{[O,\rho]}$, the two parts of this definition, ***(a)*** *and* ***(b)****, can be joined in one by using oriented segment and the condition*

$$\frac{\overline{OA'}}{\overline{OA}} = \rho \iff \overline{OA'} = \rho \cdot \overline{OA}, \quad where \quad \rho \neq 0 \quad real\ number.$$

Remarks: (a) It is an easy geometrical exercise to prove synthetically that any homothety $\mathcal{H}_{[O,\rho]}$, ($\rho \neq 0$) is a one to one and onto mapping of the plane.

(b) If $\rho = 0$, then every point $A \in \mathcal{P}$ is mapped to the point O, and so the mapping $\mathcal{H}_{[O,0]}$ is constant $\{\mathcal{H}_{[O,0]}(\mathcal{P}) = \{O\}\}$, and therefore not geometrically interesting.

Lemma 4.2.1 *If we know the ratio $\rho \neq 0$ of a homothety and a point A along with its image A' [so, (A, A') is a pair of corresponding points], then we can find the center O of the homothety.*

Proof

(a) If $A = A'$ then the center $O = A = A'$ and so the homothety is $H_{[A,\rho]}$.

(b) If $\rho = 1$, then the homothety must be either the identity and so $A = A'$, or a parallel translation by the vector $\overrightarrow{AA'}$ (when $\rho = 1$ and $A \neq A'$) and so its center is the infinity point of the straight line AA'.

(c) Otherwise $A \neq A'$ and $0 \neq \rho \neq 1$. Then, the center O is on the straight line AA', the vector $\overrightarrow{AA'} \neq \vec{0}$ is known, and we have

$$\frac{\overrightarrow{OA'}}{\overrightarrow{OA}} = \rho = \frac{\rho}{1}$$

and so,

$$\frac{\overrightarrow{OA'} - \overrightarrow{OA}}{\overrightarrow{OA}} = \frac{\rho - 1}{1}.$$

Hence,

$$\frac{\overrightarrow{AA'}}{\overrightarrow{OA}} = \frac{\rho - 1}{1}$$

and so,

$$\overrightarrow{OA} = \frac{1}{\rho - 1}\overrightarrow{AA'}.$$

This relation determines O on the line AA' (make Figures).

Similarly, or more straightforward, we find

$$\overrightarrow{OA'} = \overrightarrow{OA} + \overrightarrow{AA'} = \left(\frac{1}{\rho - 1} + 1\right)\overrightarrow{AA'} = \frac{\rho}{\rho - 1}\overrightarrow{AA'}.$$

We can also use this relation to determine O on the line AA' (make figures). ■

Circles and Homotheties.

See **Figure 4.5** and fill in the details of the proof of the **Result** stated below, as a straightforward exercise.

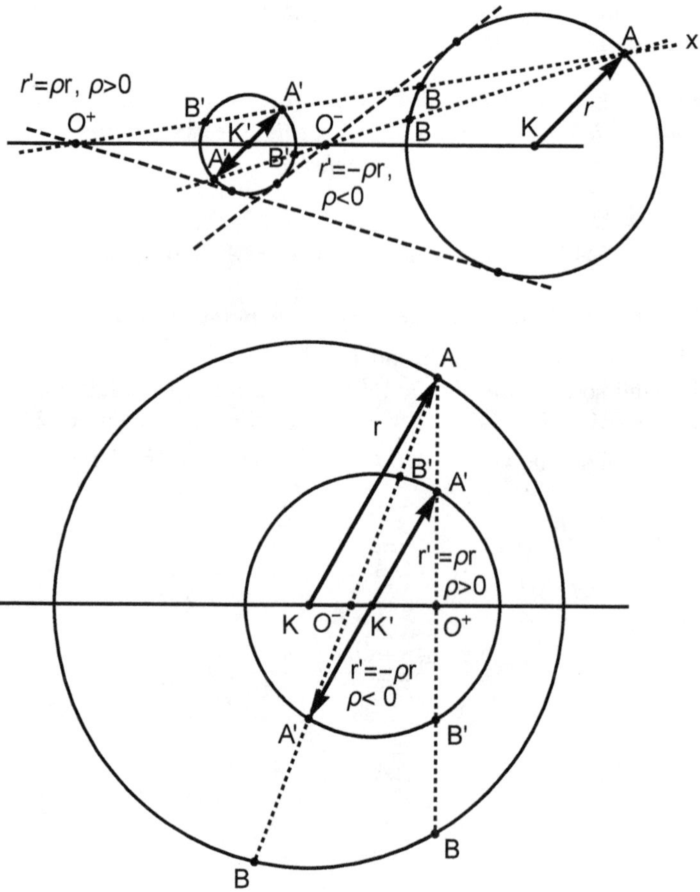

Figure 4.5: Homothetic circles. The homothetic ratio satisfies $\rho = \pm\dfrac{r'}{r} = \pm\dfrac{OK'}{OK}$ **and** $\rho = \pm\dfrac{r'}{r} = \pm\dfrac{O^{\pm}K'}{OK}$

Result: *A homothety* $H_{[O,\rho]}$ *maps a circle* $C[K, r]$ *to a circle* $C[K', r']$ *of radius*

$$r' = |\rho|r,$$

and its center K' is on the ray OK, if $\rho > 0$, and on the opposite ray, if $\rho < 0$ and satisfies the relation

$$OK' = |\rho| OK.$$

So the center K' is the homothetic image of the center K.

This result has several **consequences** some of which are stated next. Fill in the missing details in their proofs as straightforward exercises, using basic facts of the classical plane geometry and properties of similar triangles.

We consider any two corresponding vector-radii $\overrightarrow{KA}$ and $\overrightarrow{K'A'}$, such that $\mathcal{H}_{[O,\rho]}(A) = A'$ and $\mathcal{H}_{[O,\rho]}(K) = K'$. Necessarily, $\overrightarrow{KA} \parallel \overrightarrow{K'A'}$, these vectors are parallel and have:
the same direction, if

$$\rho = \frac{r'}{r} > 0;$$

opposite direction, if

$$\rho = -\frac{r'}{r} < 0.$$

Then:

(a) The center of the homothety is the point of intersection of the center straight line KK' with a straight line AA' defined by two corresponding points A and A'. If $\rho > 0$, the homothety is positive and the center O^+ is closer to the center of the smaller circle but not between K and K'. Otherwise, $\rho < 0$, the homothety is negative and the center O^- is between K and K' and closer to the center of the smaller circle.

(b) In particular, O is the intersection of KK' and the common tangents of the two circles, when common tangent(s) exist(s).

So, working backward, if we know the radii r and r' and the position of the centers K and K' of the circles $C[K,r]$ and $C[K',r']$, we can find the two homotheties that map $C[K,r]$ onto $C[K',r']$. The centers O^+ or O^- of the homotheties are on the straight line KK', and the ratios are

$$\rho = \pm\frac{r'}{r}.$$

The centers K and K' of the circles are corresponding homothetic points and must satisfy

$$OK' = |\rho| OK.$$

For convenience, we consider the line KK' as an axis and then we can compute O^+ and O^- as follows:

(1) For the positive homothety that maps the circle $C[K, r]$ to the circle $C[K', r']$, the ratio

$$\rho = \frac{r'}{r} > 0.$$

So, using **Lemma 4.2.1** and its **proof**, we find:

$$\overline{O^+K} = \frac{r}{r'-r}\overline{KK'} = \frac{r}{r-r'}\overline{K'K}$$

and

$$\overline{O^+K'} = \frac{r'}{r'-r}\overline{KK'} = \frac{r'}{r'-r}\overline{K'K}.$$

Each of these two relations determines the center O^+ of the positive homothety. (We could also use the similarity of the triangles KAO^+ and $K'A'O^+$ to prove this relation.)

(2) For the negative homothety that maps the circle $C[K, r]$ to the circle $C[K', r']$, the ratio is

$$\rho = -\frac{r'}{r} < 0.$$

So, by **Lemma 4.2.1** and its **proof**, we find:

$$\overline{O^-K} = \frac{-r}{r+r'}\overline{KK'} = \frac{r}{r+r'}\overline{K'K}$$

and

$$\overline{O^-K'} = \frac{r'}{r+r'}\overline{KK'} = \frac{-r'}{r+r'}\overline{K'K}.$$

Each of these two relations determines the center O^- of the negative homothety. (We could also use the similarity of the triangles KAO^- and $K'A'O^-$, to prove this relations.)

4.3 Power of a Point and Radical Axis

Power of a Point with respect to a Circle.

We consider a circle (C, r) and any point O of its plane. We also consider any straight line though O that intersects the circle at two points F and G. We also draw the diametrical line OC that interests the circle at two

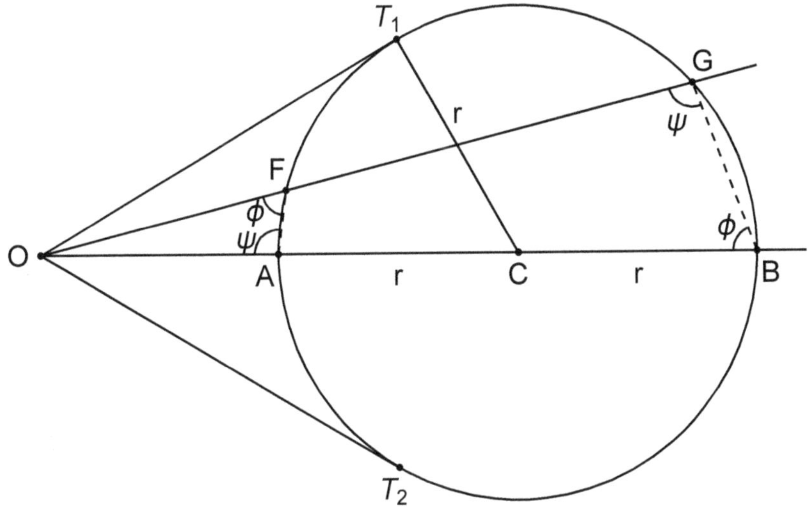

Figure 4.6: Power of $O = \overline{OF} \cdot \overline{OG} = \overline{OA} \cdot \overline{OB} = OC^2 - r^2 = OT_1^2 = OT_2^2 > 0$

points A and B. Given the circle (C, r) and the point O the line OC and the points A and B are fixed. See **Figures 4.6** and **4.7**.

Then by the inscribed quadrilateral $ABGF$, we conclude that the triangles OAF and OBG are similar. (Opposite similar.) Therefore

$$\frac{OA}{OG} = \frac{OF}{OB} \iff OF \cdot OG = OA \cdot OB,$$

which is fixed, independent of the line OFG.

In **Figure 4.6** the point O is outside the circle (C, r). We also consider the tangents OT_1 and OT_2 from O to the circle, and we have

$$OA \cdot OB = (OC - r)(OC + r) = OC^2 - r^2 = OT_1^2 = OT_2^2.$$

In **Figure 4.7** the point O is inside the circle (C, r). We have

$$OA \cdot OB = (-OC + r)(OC + r) = -OC^2 + r^2.$$

So far, the segments used were not oriented (all positive). But, if we use oriented segments along the lines OFG, we obtain:

(a) If O is outside the circle (C, r), the above fixed product is equal to the following expressions

$$\overline{OF} \cdot \overline{OG} = \overline{OA} \cdot \overline{OB} = (OC - r)(OC + r) = OC^2 - r^2 = OT_1^2 = OT_2^2 > 0.$$

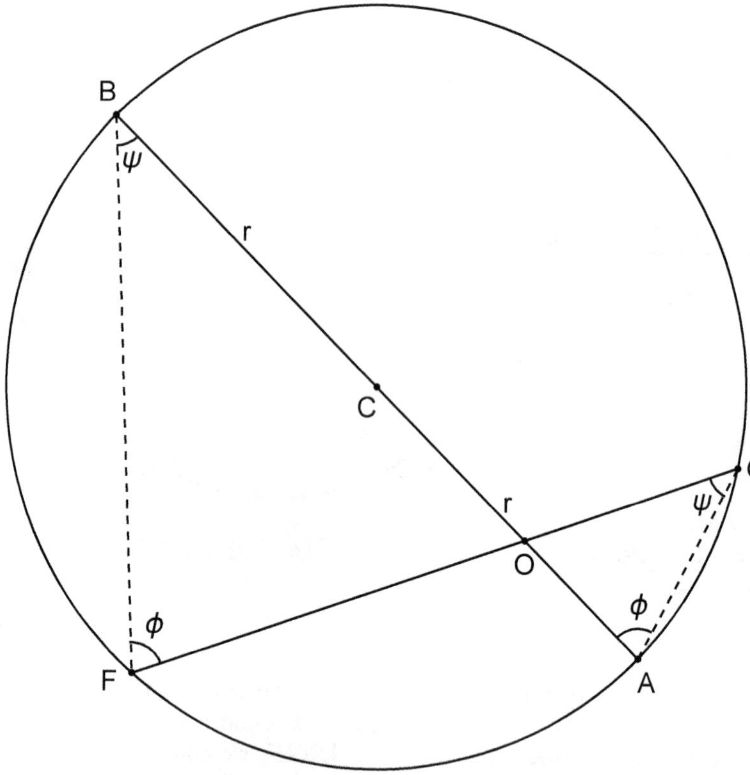

Figure 4.7: Power of $O = \overline{OF} \cdot \overline{OG} = \overline{OA} \cdot \overline{OB} = OC^2 - r^2 < 0$

(The same as before.)

(b) If O is inside the circle (C, r), the above fixed product is equal to the following expressions

$$\overline{OF} \cdot \overline{OG} = \overline{OA} \cdot \overline{OB} = (OC - r)(OC + r) = OC^2 - r^2 < 0,$$

(negative answer now.)

Hence, we define the (algebraic) **power of the point** O **with respect to the circle** $\mathcal{C} = (C, r)$ the fixed (algebraic) product $\overline{OF} \cdot \overline{OG}$. We denote this product and it is equal to:

(a) If O is **outside** the circle (C, r), then the power of O has the following expressions

$$\mathfrak{P}_C(O) := \overline{OF} \cdot \overline{OG} = \overline{OA} \cdot \overline{OB} =$$
$$(OC - r)(OC + r) = OC^2 - r^2 = OT_1^2 = OT_2^2 > 0.$$

(b) If O is **inside** the circle (C, r), then the power of O has the following expressions

$$\mathfrak{P}_C(O) := \overline{OF} \cdot \overline{OG} = \overline{OA} \cdot \overline{OB} = (OC - r)(OC + r) = OC^2 - r^2 < 0.$$

(c) If the point O is **on** the circle, let us say $O = F$, then $OF = FF$ (zero segment), and so, the power of O, in this case, is zero.

Using **analytic geometry**, we can prove the result

$$\overline{OF} \cdot \overline{OG} = OC^2 - r^2,$$

for any line OFG intersecting the circle (C, r) at the points F and G, in the following way.

Any circle in the $x - y$ plane has equation

$$x^2 + y^2 - 2ax - 2by + c = 0$$

or

$$(x - a)^2 + (y - b)^2 = a^2 + b^2 - c$$

with

$$a^2 + b^2 - c = r^2 \geq 0,$$

and so

$$(x - a)^2 + (y - b)^2 = r^2.$$

The center of circle is $C = (a, b)$ and the radius is r.

Consider any point $O = (s, t)$ of the plane at distance

$$d = \sqrt{(s - a)^2 + (t - b)^2}$$

from the center of the circle.

If we make a translation and a rotation we consider the equation of the circle and the point O to be

$$x^2 + y^2 = r^2,$$

and

$$O = (d, 0).$$

Now, any line through O and intersecting the circle at points F and G has equation

$$y = m(x - d).$$

If O is located inside the circle, then the slope m can be anything. But,if O is located outside the circle, then the slope m must satisfy

$$\frac{-r}{\sqrt{d^2 - r^2}} \leq m \leq \frac{r}{\sqrt{d^2 - r^2}}.$$

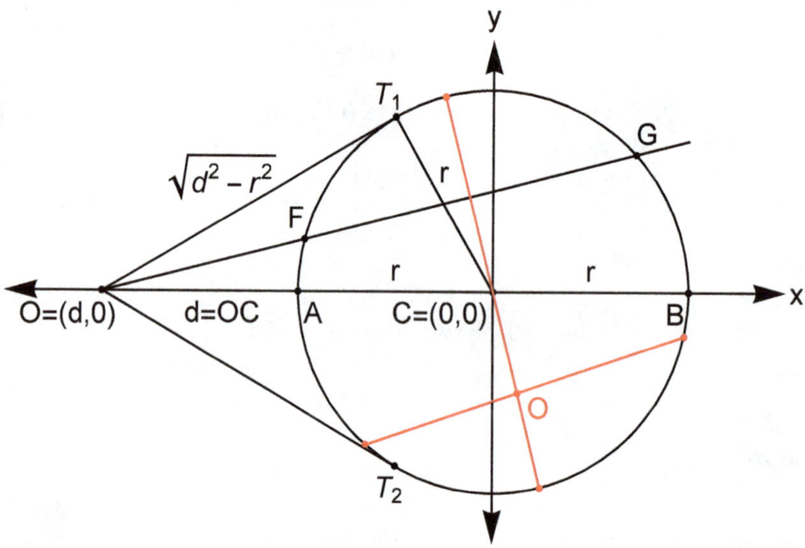

Figure 4.8: Power of $O = \overline{OF} \cdot \overline{OG} = \overline{OA} \cdot \overline{OB} = OC^2 - r^2 < 0$, with analytic geometry

Now, to find the points F and G, we must solve the system of the two equations

$$x^2 + y^2 = r^2$$
$$y = m(x - d).$$

After solving for (x, y), the solutions give us the two points

$$F = \left(\frac{dm^2 - \sqrt{-d^2m^2 + (1+m^2)r^2}}{1+m^2}, \frac{-m\left(d + \sqrt{-d^2m^2 + (1+m^2)r^2}\right)}{1+m^2} \right),$$

$$G = \left(\frac{dm^2 + \sqrt{-d^2m^2 + (1+m^2)r^2}}{1+m^2}, \frac{m\left(-d + \sqrt{-d^2m^2 + (1+m^2)r^2}\right)}{1+m^2} \right).$$

We now compute $d_1 :=$ distance(O, F) and $d_2 :=$ distance(O, G), and we find

$$d_1 = \sqrt{\frac{-d^2(-1+m^2) + (1+m^2)r^2 + 2d\sqrt{-d^2m^2 + (1+m^2)r^2}}{1+m^2}},$$

$$d_2 = \sqrt{\frac{-d^2(-1+m^2) + (1+m^2)r^2 - 2d\sqrt{-d^2m^2 + (1+m^2)r^2}}{1+m^2}}.$$

Then, the product of the oriented distances is

$$\overline{d_1} \cdot \overline{d_2} = \pm\sqrt{(d^2 - r^2)^2} = \pm\left|d^2 - r^2\right| = \begin{cases} d^2 - r^2 > 0, & \text{if} \quad d > r, \\ 0, & \text{if} \quad d = r, \\ d^2 - r^2 < 0, & \text{if} \quad d < r. \end{cases}$$

The two points F and G coincide, when $|d| > r$ and the intersecting line(s) become(s) tangent(s) to the circle. This happens, if m is such that,

$$m = \frac{\pm r}{\sqrt{d^2 - r^2}} \iff m^2d^2 = (1+m^2)r^2 \iff$$

$$F = G = \left(\frac{dm^2}{1+m^2}, \frac{-dm}{1+m^2} \right) = \left(\frac{r^2}{d}, \frac{\mp r\sqrt{d^2 - r^2}}{d} \right).$$

The above results agree with result found before, synthetically.

Radical Axis of Two Coplanar Circles.

Now we are going to study the **radical axis of two coplanar circles** $\mathcal{C}_1 = C[O_1, R_1]$ and $\mathcal{C}_2 = C[O_2, R_2]$ and its properties. We have used the radical axis and its properties in the preceding material and the associated Poncelet points.

By **definition**, ***"the radical axis of two given coplanar circles*** *is the geometrical locus of the points of their plane that have equal powers with respect to these circles".*

We assume that $O_1 \neq O_2$ (the circles are not concentric). Suppose we have a point of the plane $K \in \mathcal{P}$, such that $\mathfrak{P}_{\mathcal{C}_1}(K) = \mathfrak{P}_{\mathcal{C}_2}(K)$ and consider $d_1 = KO_1$ and $d_2 = KO_2$. See **Figure 4.9.**

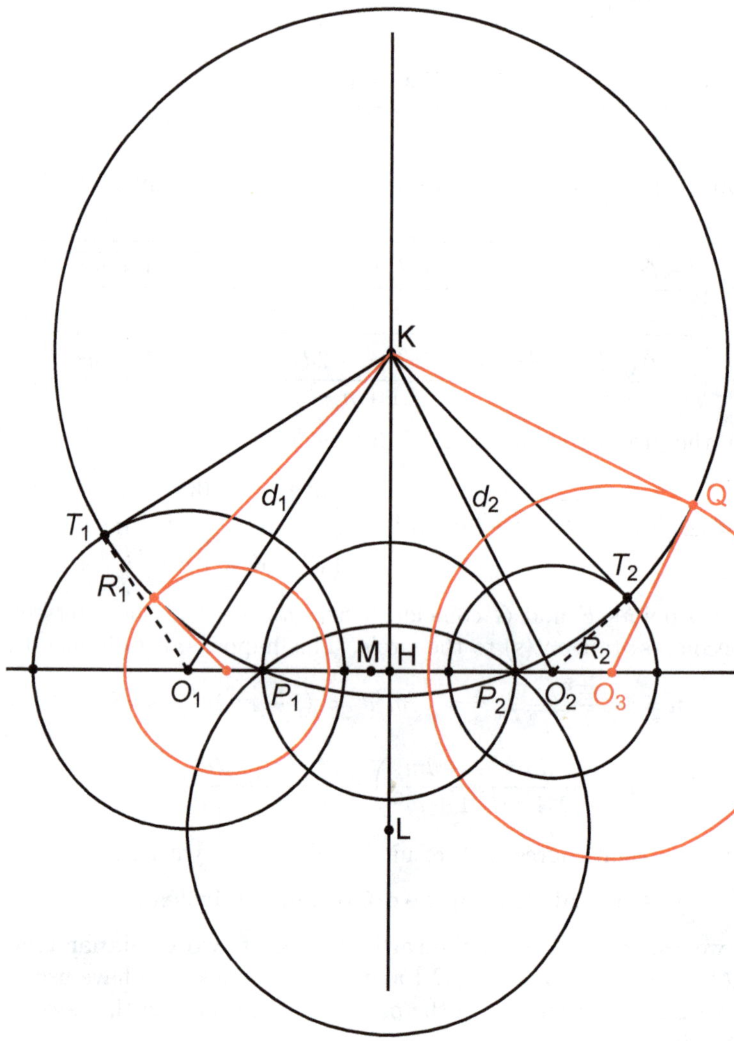

Figure 4.9: Radical Axis of Two Circles

Since,

$$\mathfrak{P}_{\mathcal{C}_1}(K) = KO_1^2 - R_1^2 = d_1^2 - R_1^2, \qquad \mathfrak{P}_{\mathcal{C}_2}(K) = KO_2^2 - R_2^2 = d_2^2 - R_2^2,$$
$$\text{and} \quad \mathfrak{P}_{\mathcal{C}_1}(K) = \mathfrak{P}_{\mathcal{C}_2}(K), \quad K \text{ must satisfy} \quad d_1^2 - d_2^2 = R_1^2 - R_2^2.$$

Then, by relation **(b)** of the **Theorem of the Medians, 4.1.1** [or **(b)** of its **Corollary**], applied to the triangle KO_1O_2, we obtain

$$2\overline{O_1O_2} \cdot \overline{MH} = R_1^2 - R_2^2 \iff \overline{MH} = \frac{R_1^2 - R_2^2}{2\overline{O_1O_2}}, \tag{4.2}$$

where M is the midpoint of O_1O_2 (known point) and H is the projection of K onto O_1O_2. Therefore, H is a fixed point on O_1O_2, predetermined by the above relation, and the point K is on the straight line perpendicular to O_1O_2 at the point H.

Now, using the **Theorem of the Medians, 4.1.1, (b)**, we directly prove that any point of the straight line HK satisfies the definition of radical axis and no other point of the plane does. (Provide details.) Therefore:

"The ***radical axis*** *of $\mathcal{C}_1 = C[O_1, R_1]$ and $\mathcal{C}_2 = C[O_2, R_2]$ is the whole straight line perpendicular to the center line O_1O_2 at the point H (in equation 4.2)."*

Remarks and Further Results: Now we can justify the following:

(a) If $O_1 \neq O_2$ and $R_1 = R_2$, then $H = M$ (the midpoint of the center segment O_1O_2).

(b) If $O_1 = O_2$ and $R_1 \neq R_2$, then H is the infinity point of the straight line O_1O_2 and the radical axis is the line at infinity.

(c) If $O_1 = O_2$ and $R_1 = R_2$, then the two circles coincide and it makes no sense to ask for the radical axis of one circle. (In this case, every point of the plane satisfies the definition of the radical axis.)

(d) If the circles are tangent (internally or externally), then $O_1O_2 = |R_1 \pm R_2|$ and the radical axis is the common tangent.

(e) If the circles intersect at two points, then $|R_1 - R_2| < O_1O_2 < R_1 + R_2$ and the radical axis is the straight line of the common chord. (The common points have power zero, the points of the cord inside the circles have equal negative powers, and the points outside the circles have positive equal powers.)

(f) If the circles are outside each other, then $O_1O_2 > R_1 + R_2$ and the radical axis is located between them and does not intersect them (**Figure 4.9**). The point H lies outside both circles and between the midpoint M and the center of the smaller circle.

Proof First of all for all positions of the two given circles, examined above or below, using oriented segments (lengths) (even if one radius or both radii is / are zero), we have the following two useful equations

$$\overline{O_1H} = \overline{O_1M} + \overline{MH} = \frac{\overline{O_1O_2}}{2} + \frac{R_1^2 - R_2^2}{2\overline{O_1O_2}} = \frac{O_1O_2^2 + R_1^2 - R_2^2}{2\overline{O_1O_2}},$$

$$\overline{O_2H} = \overline{O_2M} + \overline{MH} = -\frac{\overline{O_1O_2}}{2} + \frac{R_1^2 - R_2^2}{2\overline{O_1O_2}} = \frac{-O_1O_2^2 + R_1^2 - R_2^2}{2\overline{O_1O_2}}.$$

(See also **Equation 3.8**.) Let us consider the direction from O_1 to O_2 to be the positive direction. Then $\overline{O_1H}$ is positive and $\overline{O_2H}$ is negative. Since the circles are outside each other, we have $R_1 + R_2 < O_1O_2$. Then using the above equations and this inequality, we prove that

$$\overline{O_1H} > R_1 \quad \text{and} \quad |\overline{O_2H}| = \overline{HO_2} > R_2.$$

(Analyzing the inequalities and using $R_1 + R_2 < O_1O_2$, we reach a final consistent inequality. Therefore these inequalities are valid. Work it out.) This proves that H lies outside both circles.

(g) If the (non-concentric) circles are inside each other and not concentric, then $0 < O_1O_2 < |R_1 - R_2|$ and the radical axis is located outside both of them and on the side that the circles are closer to each other. (**Figures 3.2, 3.3.**)

The **proof** of this item is analogous to the proof of (f) above.

(h) The tangent segments to the two circles from any point K of the part of radical axis outside the intersection of the two circles are equal.

(i) The circles with centers the points of the part of radical axis outside the intersection of the two circles and radii the tangent segments to the circles from these points are orthogonal to the two given circles, and so the given circles are orthogonal to them.

(j) (I) If the circles do not intersect, then $O_1O_2 > R_1 + R_2$ (they are outside each other) or $|R_1 - R_2| > O_1O_2$ (one circle is inside the other one), and all the circles with centers on the radical axis of $\mathcal{C}_1 = C[O_1, R_1]$ and

$\mathcal{C}_2 = C[O_2, R_2]$ and orthogonal to them are concurrent at two points P_1 and P_2 on the center line O_1O_2, (as in **Figure 4.9**).

These points are called the **Poncelet points** of $\mathcal{C}_1 = C[O_1, R_1]$ and $\mathcal{C}_2 = C[O_2, R_2]$. {Poncelet named them **limit points** of the family of circles with centers in the center line O_1O_2, of $\mathcal{C}_1 = C[O_1, R_1]$ and $\mathcal{C}_2 = C[O_2, R_2]$, and with radical axis the radical axis of $\mathcal{C}_1 = C[O_1, R_1]$ and $\mathcal{C}_2 = C[O_2, R_2]$. All these circles have the same Poncelet points and the Poncelet points are the smallest circles of this family, (point circles). They are the two real roots of **equation 3.8**.}

(II) If the circles $\mathcal{C}_1 = C[O_1, R_1]$ and $\mathcal{C}_2 = C[O_2, R_2]$ are tangent, then $O_1O_2 = R_1 + R_2$, the Poncelet points coincide at their point of tangency. (The double root of **equation 3.8**.)

(III) If the circles $\mathcal{C}_1 = C[O_1, R_1]$ and $\mathcal{C}_2 = C[O_2, R_2]$ intersect, then $O_1O_2 < R_1 + R_2$, the Poncelet points are the two common points. (The two complex conjugate roots of **equation 3.8**).

Proof The cases (II) and (III) are rather obvious, and so we will study case (I).

Proof of (I): We consider any circle $C[K, r]$ with center a point K on the radical axis and orthogonal to both $C[O_1, R_1]$ and $C[O_2, R_2]$. Its radius is $r := KT_1 = KT_2$, where KT_1 and KT_2 are the two equal tangent segments from K to the circles $C[O_1, R_1]$ and $C[O_2, R_2]$, respectively.

Since,

$$R_1^2 + KT_1^2 = O_1K^2 = O_1H^2 + KH^2$$

and the circles are disjoint, H lies outside both circles, i.e., $O_1H > R_1$ and $O_2H > R_2$, we have that

$$r := KT_1 = KT_2 > KH.$$

This inequality implies that any circle $C[K, r]$ intersects the center line O_1O_2 at two points. Let us call them P_1 and P_2.

Therefore, P_1P_2 is a chord of any circle $C[K, r]$. Since $KH \perp P_1P_2$, H is the midpoint of P_1P_2, or $\overline{P_1H} = \overline{HP_2}$. Also, O_1 and O_2 are outside the circles $C[K, r]$.

The power of O_1 with respect to the circle $C[K, r]$ is $O_1T_1^2 = \overline{O_1P_1} \cdot \overline{O_1P_2}$, and since $O_1T_1 = R_1$ we find,

$$R_1^2 = (\overline{O_1H} - \overline{HP_1}) \cdot (\overline{O_1H} + \overline{HP_2}) =$$
$$(\overline{O_1H} - \overline{HP_1}) \cdot (\overline{O_1H} + \overline{HP_1}) = O_1H^2 - HP_1^2.$$

Therefore,

$$HP_1^2 = O_1H^2 - R_1^2, \quad (\Longleftrightarrow \quad O_1H^2 = HP_1^2 + R_1^2 > R_1^2).$$

Similarly, $$HP_2^2 = O_2H^2 - R_2^2, \quad (\Longleftrightarrow \quad O_2H^2 = HP_2^2 + R_2^2 > R_2^2).$$

So, H is outside both circles $C[O_1, R_1]$ and $C[O_2, R_2]$, and since $\overline{P_1H} = \overline{HP_2}$, we have the important relation, for the absolute values of the respected oriented segments

$$\sqrt{O_1H^2 - R_1^2} = HP_1 = \frac{P_1P_2}{2} = HP_2 = \sqrt{O_2H^2 - R_2^2}.$$

Thus, the points P_1 and P_2 are predetermined by R_1, R_2, O_1O_2, and its midpoint M. So, P_1 and P_2 are the same for any circle $C[K, r]$. Also, fact **(4)** follows immediately from this equation.

Using the above relations, we easily prove that: When the initial circles $C[O_1, R_1]$ and $C[O_2, R_2]$ are outside of each other, then P_1 is inside one of them and P_2 inside the other one. Otherwise, one of these points is inside both of them and the other outside. (**Figures 3.2, 3.3**. See also **equation 3.8**.)

(k) The smallest circle orthogonal to $\mathcal{C}_1 = C[O_1, R_1]$ and $\mathcal{C}_2 = C[O_2, R_2]$ is the circle with diameter P_1P_2 and center the point H (midpoint of P_1P_2), (as in **Figure 4.9**). This circle is called the **Poncelet circle** of $\mathcal{C}_1 = C[O_1, R_1]$ and $\mathcal{C}_2 = C[O_2, R_2]$. (See also **Figures 3.2, 3.3**.)

(l) Suppose we pick a point Q on any circle $C[K, r]$ (so, $r = KQ$), and draw the tangent to this circle at this point. This tangent intersects the center line O_1O_2 at a point O_3. Then, the circle $C[O_3, O_3Q]$ is orthogonal to all the circles with centers on the radical axis and orthogonal to $\mathcal{C}_1 = C[O_1, R_1]$ and $\mathcal{C}_2 = C[O_2, R_2]$. In this way we can construct all such circles, as $C[O_3, O_3Q]$. All such circles have radical axis HK and the Poncelet points P_1 and P_2, the same as with $\mathcal{C}_1 = C[O_1, R_1]$ and $\mathcal{C}_2 = C[O_2, R_2]$. (This follows from all the above facts.)

(m) Here, we have two families (pencils) of circles orthogonal to each other:

1st *family:* All the circles with centers on the radical axis of $\mathcal{C}_1 = C[O_1, R_1]$ and $\mathcal{C}_2 = C[O_2, R_2]$ and orthogonal to them.

2nd *family:* The circles with centers on the center line O_1O_2 and orthogonal to the first family.

The radical axis of the second family is the center line O_1O_2. The points P_1 and P_2 are the Poncelet points of both families. (This follows from all the above facts.)

(n) Important Property of the Poncelet Points.

The existence of the Poncelet points has an important consequence. All the circles $C[K, r]$ with center on the radical axis of $C[O_1, R_1]$ and $C[O_2, R_2]$ and orthogonal to them, pass through both Poncelet points. Since inversions preserve the absolute value of angles, if we apply an inversion with center one of the Poncelet points and any power, the circles $C[K, r]$ transform to straight lines orthogonal to the images of the circles $C[O_1, R_1]$ and $C[O_2, R_2]$ and pass through the image point of the other Poncelet point. Therefore, these infinitely many straight lines are common diameters to both image-circles. This implies that the two image-circles are concentric with center the inverse image of the other Poncelet point. See **Figure 4.10**. So, we have proven:

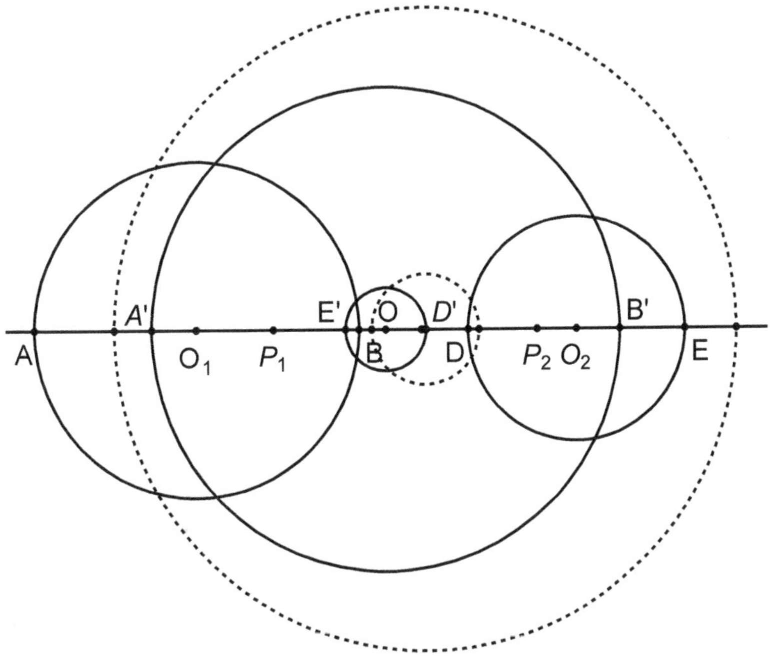

Figure 4.10:
Two non-intersecting circles mapped to concentric circles by an appropriate inversion

Theorem 4.3.1 *Given any two disjoint circles, an inversion with center one of the two Poncelet points and any power, maps the two given circles* *onto two concentric circles with center the image of the other Poncelet point. Similarly, all circles that do not intersect and have the same radical axis can be mapped to concentric circles by this inversion.*

In **Figure 4.10**, the circles $C[O_1, R_1]$ and $C[O_2, R_2]$ are mapped to two concentric circles with center O by an inversion having center the Poncelet point P_1 and some positive power. The circle $C[O_1, R_1]$ is mapped to the large circle and the circle $C[O_2, R_2]$ to the smaller circle. The other Poncelet point, P_2, is mapped to O. The points A and B are mapped to A' and B'. The dotted circles are obtained by using an inversion with center the Poncelet point P_2. The points D and E are mapped to D' and E'.

The particular choices of these points, in **Figure 4.10**, make the segments $A'B'$ and $D'E'$ have common midpoint the point O.

(This result, of inverting two non-intersecting circles onto two concentric circles, can be obtained algebraically, but the system of equations obtained is rather cumbersome to solve. So, we have provided a direct synthetic geometric solution to the problem.)

4.4 Inversion

The transformation of inversion in the Euclidean plane $\mathcal{P}$ is defined as follows.

Definition 4.4.1 *We consider any point $O \in \mathcal{P}$ and any real number $c \neq 0$. We define the inversion with respect to the data O and c to be the mapping*

$$\mathfrak{I}_{[O,c]} \;:\; \mathcal{P} \longrightarrow \mathcal{P} \cup \{\infty\} \quad \text{such that:}$$

(a) $\forall\ P \in \mathcal{P}$ and $P \neq O$, then $\mathfrak{I}_{[O,c]}(P) = P'$, if the following two conditions are met:
(1) O, P and P' are on the same line (collinear points), i.e., P' is located on the line OP, and
(2)

$$\overline{OP} \cdot \overline{OP'} = c \iff \overline{OP'} = \frac{c}{\overline{OP}} \left(\iff \overline{OP} = \frac{c}{\overline{OP'}} \right).$$

(b) If $P = O$, then $\overline{OP} = 0$ and we define $\mathfrak{I}_{[O,c]}(O) = \infty$.

In relation to this definition, we also use the terms:
(a) The point O is called the **center** or the **pole of the inversion.**
(b) The number c is called the **power of the inversion.**

Figure 4.11: Positive (or direct) inversion
$\overline{OP} \cdot \overline{OP'} = c > 0$

(c) If the power $c > 0$, then the point P' is on the ray $\overrightarrow{OP}$, and the inversion is called **positive** or **direct**. See **Figure 4.11**.

(d) If the power $c < 0$, then the point P' is on the ray opposite to $\overrightarrow{OP}$, hence O is located between P and P' on the line PP', and the inversion is called **negative**, or **opposite**, or **indirect**. See **Figure 4.12**.

Figure 4.12: Negative (or Opposite, or Indirect) inversion
$\overline{OP} \cdot \overline{OP'} = c < 0$

The negative inversion is equal to the commutative composition of the corresponding positive inversion and a half-turn of the same center. That is, in general,

$$\mathfrak{I}_{[O,c]} = \mathfrak{I}_{[O,-c]} \circ \mathcal{R}_{[O,\pi]} = \mathcal{R}_{[O,\pi]} \circ \mathfrak{I}_{[O,-c]}.$$

From the definition of inversion, we immediately gather two properties:
(1) The composition of an inversion $\mathfrak{I}$ with itself is equal to the identity, $\mathfrak{I} \circ \mathfrak{I} = I$. Or, any inversion is the inverse of itself, $\mathfrak{I}^{-1} = \mathfrak{I}$. This property says that any inversion is an **involutary transformation**.
(2) If $(A,\ A')$ and $(B,\ B')$ are two pairs of inverse points, then either the five points $O, A,\ A',\ B,\ B'$ are collinear (where O is the center of the inversion), or the four points $A,\ A',\ B,\ B'$ are concyclic.

Next, we consider an **inversion** $\mathfrak{I}$ with center O and power c and a **circle** $C[K,\rho]$ and we want to find its inversive image $\mathfrak{I}\{C[K,\rho]\}$.

In the sequel, we let $p = \mathfrak{P}_{C[K,\rho]}(O)$, the power of the center of inversion O with respect to the circle $C[K,\rho]$.

If $c > 0$, the circle of inversion $C[O,\sqrt{c}]$ is point-wise invariant under $\mathfrak{I}$, and any circle orthogonal to the circle of inversion $C[O,\sqrt{c}]$ is mapped to itself by the inversion (not point-wise, but set-wise.)
If $c < 0$, the circle of inversion and any circle pseudo-orthogonal to it is mapped to itself by the inversion (set-wise, not point-wise). (A circle $\mathcal{D}$ is **pseudo-orthogonal** to a circle $\mathcal{C}$, if $\mathcal{D}$ intersects $\mathcal{D}$ at diametrical points

of $\mathcal{C}$. The property of pseudo-orthogonality, unlike the property of orthogonality, is not symmetrical.)
So, in these two cases we know the answer to the above question, and $p = c$.
If $C[K, \rho]$ passes through O, then its inversive image $\mathfrak{I}\{C[K, \rho]\}$ is the straight line perpendicular to OK at the point P, such that $\overline{OP} \cdot \overline{OP'} = c$, where P' is the point of $C[K, \rho]$, diametrical to O. (See bibliography.)

Here, we will study the case that we have used a lot of the inverses of the circles that do not contain O and list the results that follow. To this end, we will use **Figures 4.13** and **4.14**. Since $O \notin C[K, \rho]$, it holds $p \neq 0$.

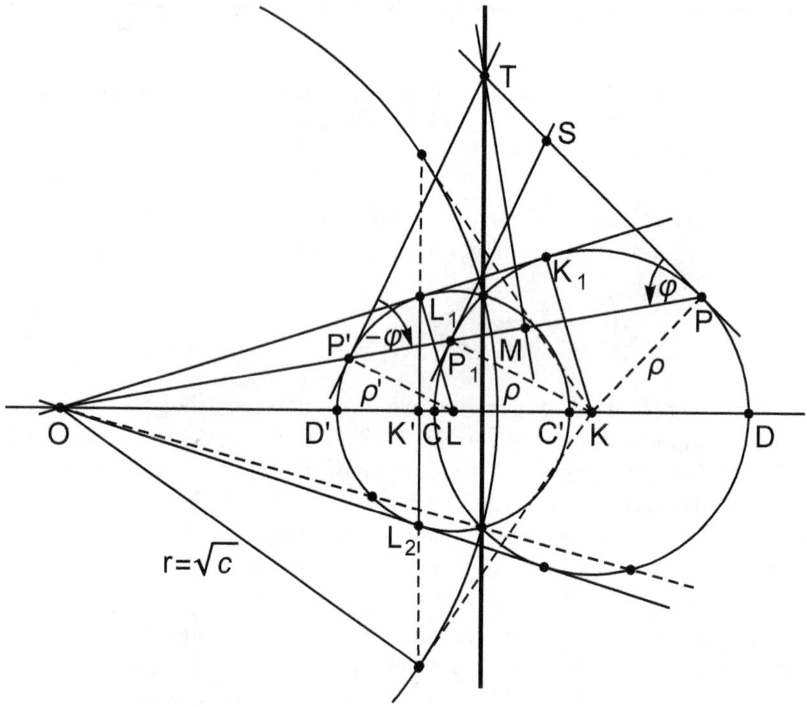

Figure 4.13: Inverse of a circle not passing through O. ($c > 0$ and some properties)

We pick any point $P \in C[K, \rho]$ and we consider its inverse P'. We also consider the second point P_1 at which the line OP intersects the circle

$C[K,\rho]$. In general, $P \neq P_1$, unless OP is tangent to the circle, in which case $P = P_1$. See **Figures 4.13** and **4.14**.

Then we have,

$$\overline{OP} \cdot \overline{OP'} = c$$

and

$$\overline{OP} \cdot \overline{OP_1} = p.$$

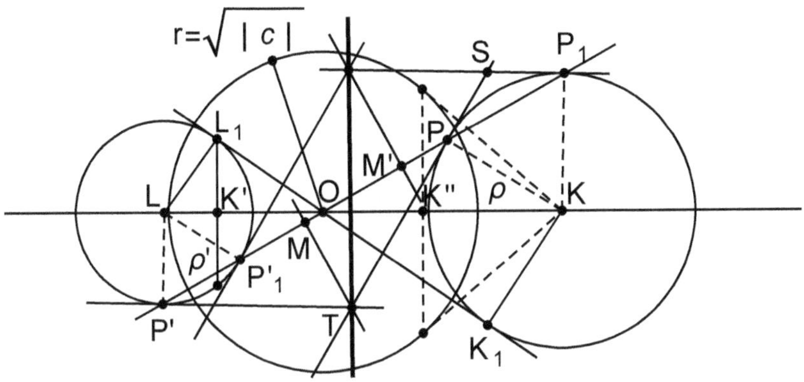

Figure 4.14: Inverse of a circle not passing through O. ($c < 0$ and some properties)

Dividing the two equations, we find

$$\frac{\overline{OP'}}{\overline{OP_1}} = \frac{c}{p}.$$

This means that P_1 corresponds to P' under the homothety $\mathcal{H}$ with homothetic center O, the center of inversion, and homothetic ratio

$$\lambda = \frac{c}{p}.$$

As P moves along the given circle $C[K,\rho]$, the point P_1 traces the same circle and the point P', which is the inverse point of P, traces the homothetic circle

$$C[L,\rho'] = \mathcal{H}\{C[K,\rho]\}.$$

So, we have the following Theorem:

Theorem 4.4.1 *The image of a circle $C[K,\rho]$ under an inversion $\mathfrak{I}_{[O,c]}$, such that the center of inversion O is not on this circle, is the circle $C[L,\rho']$,* *which is the homothetic image of $C[K,\rho]$, under the homothety $\mathcal{H}_{[O,\frac{c}{p}]}$, where c is the power of inversion and*

$$p = \mathfrak{P}_{C[K,\rho]}(O) = OK^2 - \rho^2$$

is the power of the center of inversion, O, with respect to the given circle $C[K,\rho]$.

{*That is: the homothety $\mathcal{H}_{[O,\frac{c}{p}]}$ has center the center O of the given inversion $\mathfrak{I}_{[O,c]}$, and ratio*

$$\lambda = \frac{c}{p},$$

where

$$p = \mathfrak{P}_{C[K,\rho]}(O) = OK^2 - \rho^2.$$

So,

$$\overline{OL} = \frac{c}{p} \cdot \overline{OK},$$

$$\rho' = \frac{|c|}{|p|} \cdot \rho,$$

and

$$\mathcal{H}_{[O,\frac{c}{p}]}\{C[K,\rho]\} = \mathfrak{I}_{[O,c]}\{C[K,\rho]\} := C[L,\rho'].$$

Attention: The center L is homothetic to the center K, but not inverse. Hence, L, K, and the center of inversion O, which is also the center of the homothety, are collinear.}

Important Remarks:

(a) If external tangents of the given circle $C[K,\rho]$ and its inverse $C[L,\rho']$ exist, then they pass through the common center O of the inversion $\mathfrak{I}_{[O,c]}$ and the homothety $\mathcal{H}_{[O,\frac{c}{p}]}$ on the center line KL. If internal tangents exist, then they pass through the common center of the negative inversion and negative homothety on the center line KL. (Common tangents do not exist when one circle is inside the other one, etc.)

(b) The points P_1 and P' are homothetic, whereas P and P' are inverse. In general $P_1 \neq P$, except for the two points of the common tangents of the $C[K,\rho]$ and $C[L,\rho']$ that pass through O. We say that the points P_1 and P' are homologous and the points P and P' are anti-homologous, with respect to homothetic center O.

(c) If the point P is on a common tangent from O to the circles $C[K,\rho]$ and $C[L,\rho']$, then $P_1 = P$ and $P' = P_1'$ is on the same tangent and $C[L,\rho']$. In this case, P' is both homothetic and inverse of P.

(d) By the properties of homotheties, the center L of the circle $C[L,\rho']$ is determined on the line OK by the equation

$$\overline{OL} = \frac{c}{p} \cdot \overline{OK},$$

and the radius is given by

$$\rho' = \frac{|c|}{|p|} \cdot \rho.$$

(e) Assume that the circles have common tangents. **The inverse point K' of the center K is not L, but a point $K' \neq L$ on the straight line KLO. To find K',** we draw L_1K_1, one of the common tangents to the two circles. From the properties of homotheties, we know that L_1K_1 passes through O.

Then,

$$OK \cdot OK' = c = OK_1 \cdot OL_1.$$

Therefore, the quadrilateral $K'KK_1L_1$ is inscribable and so $\widehat{KK'L_1} = \frac{\pi}{2}$, (because $\widehat{KK_1L_1} = \frac{\pi}{2}$). Hence, $L_1K' \perp OL$. Since OL_1 is tangent to the circle $C[L,\rho']$ at L_1, the triangle L_1LO is right with hypothenuse OL and corresponding altitude L_1K'. Therefore, $OK' < OL$, and

$$\text{if} \quad c > 0 \quad \text{and} \quad p > 0, \quad \text{then} \quad \overline{OK'} = \frac{c}{\overline{OK}} < \overline{OL} = \frac{c}{p} \cdot \overline{OK}.$$

{More general, with any $c (\neq 0)$, if $p > 0$, then

$$OK' = \frac{|c|}{OK} < OL = \frac{|c|}{p} \cdot OK.$$

(I) If $p > 0$, this fact could be á priori determined, since in that case we have: $0 < p = OK^2 - \rho^2 < OK^2$.

(II) If $p < 0$ ($\Longleftrightarrow O$ is inside $C[K,\rho]$), make two figures analogous to **Figures 4.13** and **4.14** and compare the differences.

(III) If $p = 0$ ($\Longleftrightarrow O \in C[K,\rho]$).

(IV) Analogous work must be done when one circle is inside the other one and so common tangents do not exist.}

(f) The inverse point of O with regard to the circle $C[L, \rho']$ is K'. That is,

$$\overline{LO} \cdot \overline{LK'} = \rho'^{\,2}.$$

[This follows form the fact that the triangle LL_1O is right $\left(\widehat{L_1} = \frac{\pi}{2}\right)$, L_1K' is the height on the hypotenuse OL and $LL_1 = \rho'$.]

If $C[L, \rho']$ is replaced by a straight line l, then the inverse of O is its reflection in l.

(g) Geometrically, we can construct $C[L, \rho']$, the inverse circle of $C[K, \rho]$, in the following steps.

(I) We pick any point $P \in C[K, \rho]$ and we find its inverse point P' on OP. We also let P_1 be the other point of intersection of OP with $C[K, \rho]$.

(II) We draw the line $k \parallel KP_1$ through P'. This intersects the line OK at the point L. Now: The point L, thus determined, is the center of $C[L, \rho']$, whose radius is $\rho' = LP'$.

(h) The tangents at two inverse points P and P' of two inverse circles are symmetrical with respect to the perpendicular bisector MT of the segment PP'. See **Figures 4.13** and **4.14**.

Let T be the point of intersection of these tangents. We draw the tangent of $C[K, \rho]$ at P_1 and let S be the point of its intersection with the tangent at P. This is parallel to the tangent at P', because the corresponding radii are parallel ($KP_1 \parallel LP'$). Since the triangle SPP_1 is isosceles, so is the triangle TPP'. Now, the claim follows and moreover MT is the bisector of the angle $\widehat{PTP'}$. Therefore, these tangents form opposite angles with the straight line $OP'P$. That is, $\measuredangle(PT, PP') = -\measuredangle(P'T, P'P)$.

(i) The line OK intersects both circles $C[K, \rho]$ and $C[L, \rho']$ at diameters CD and $C'D'$. As in **Figure 4.26**, we get that C' is the inverse of C and D' the inverse of D. So, by the homothety $\mathcal{H}_{[O, \frac{c}{p}]}$, the lengths satisfy

$$C'D' = \left|\frac{c}{p}\right| CD.$$

(j) Observe that the inverse homothety

$$\mathcal{H}_{[O,\frac{1}{\lambda}]} = \mathcal{H}_{[O,\frac{p}{c}]} = \mathcal{H}_{[O,\frac{c}{p'}]},$$

where

$$p' = \mathfrak{P}_{C[L,\rho']}(O),$$

maps $C[L,\rho']$ to $C[K,\rho]$ and we also get $pp' = c^2$.

Or, we can argue with the following useful computations of two related powers of O. The power of O with respect to $C[K,\rho]$ is

$$p = OC \cdot OD = OK^2 - \rho^2 = OK_1^2 = OP \cdot OP_1 = \text{ etc.}$$

The power $p' = \mathfrak{P}_{C[L,\rho']}(O)$, of O with respect to the circle $C[L,\rho']$, is

$$p' = \overline{OC'} \cdot \overline{OD'} = \frac{c}{\overline{OC}} \cdot \frac{c}{\overline{OD}} = \frac{c^2}{p} \quad \Longrightarrow \quad pp' = c^2 \quad \Longrightarrow \quad c = \pm\sqrt{pp'}.$$

In this context, we also have the following Theorem (referred in **Section 3.9**).

Theorem 4.4.2 *Given two circles, there are infinitely many pairs of inversions of the same center lying on their center line that invert the two given circles into equal circles and the centers of these equal circles may be chosen to be any points of the center line except for the points of its intersection with the two given circles. We could also choose the two centers to coincide and thus make the equal circles identical.*

Proof Let $C[K_1,r_1]$ and $C[K_2,r_2]$ be the two given circles. If two inversions $\mathfrak{I}_{[O,c_1]}$ and $\mathfrak{I}_{[O,c_2]}$ with common center O lying on the line K_1K_2, map these circles onto two circles of equal radius (and so to equal circles), then we have that the equal radius is

$$r := r_1 \left|\frac{c_1}{p_1}\right| = r_2 \left|\frac{c_2}{p_2}\right|,$$

where p_1 and p_2 is the power of O with respect to $C[K_1,r_1]$ and $C[K_2,r_2]$, respectively.

The radii r_1 and r_2 are known. Having picked any O on the center line K_1K_2 (except for the points of intersection of the two circles and the center line), the powers p_1 and p_2 become known. Hence, we pick any c_1 and c_2 that satisfy

$$\frac{c_1}{c_2} = \pm\frac{p_1 r_2}{p_2 r_1}.$$

Then, the circles of the two inversions are $C[O, \sqrt{|c_1|}]$ and $C[O, \sqrt{|c_2|}]$. Also, the common radius of the image circles is found by the formula for r, above.

Therefore, there are infinitely many pairs of inversions of the same center O on the center line K_1K_2 that can invert the two given circles into equal circles with centers on the center line K_1K_2 of the given circles.

If $K_1 = K_2$, the two given circles are concentric. Then, in the place of center line, we can pick any line l through $K_1 = K_2$.

To choose the centers of the two equal circles to be the same point L, then we must satisfy

$$\overline{OL} = \frac{c_1}{p_1}\overline{OK_1} = \frac{c_2}{p_2}\overline{OK_2}.$$

So using this equation, we must choose O to satisfy the *system of equations*

$$\begin{aligned} p_1 &= OK_1^2 - r_1^2, \\ p_2 &= OK_2^2 - r_2^2, \\ \frac{\overline{OK_1}}{\overline{OK_2}} &= \pm\frac{p_1c_2}{p_2c_1} = \pm\frac{r_1}{r_2}. \end{aligned}$$

Then O is determined by the last equation. We may have two choices depending on the $\pm$. Next, we find p_1 and p_2 from the first two equations and subsequently we choose any compatible c_1 and c_2 (infinite choices).

(Also, when $K_1 = K_2$, then another choice is $L = K_1 = K_2$ and we can consider the inversion $\mathfrak{I}_{[L,r_1r_2]}$ coupled with the inversion $\mathfrak{I}_{[L,r_1^2]}$ or the inversion $\mathfrak{I}_{[L,r_2^2]}$. In any case, we have infinite choices of the powers of these inversions. Find them all!)

■

Remarks: (a) In the above proof, the last equation of the *system* was expected. As we have seen in the first part of **this Subsection**, inverse circles are homothetic. By the properties of the homotheties, we have

$$\frac{\overline{OK_1}}{\overline{OL}} = \pm\frac{r_1}{r} \quad \text{and} \quad \frac{\overline{OK_2}}{\overline{OL}} = \pm\frac{r_2}{r}.$$

By dividing these equalities, we obtain the equation

$$\frac{\overline{OK_1}}{\overline{OK_2}} = \pm\frac{r_1}{r_2}.$$

(b) An interesting problem is to find the inversions that each one of them inverts two given circles onto two equal circles.

Derivation of Equation 3.8.

It is interesting to prove why **equation 3.8** locates the Poncelet points of two circles $C[O_1, r_1]$ and $C[O_2, r_2]$. This is done in the following 3 steps:
Step 1:

Theorem 4.4.3 *Given two circles in the plane, the fixed points of the composition of the inversions*

$$\mathfrak{I}_{[O_2,r_2^2]} \circ \mathfrak{I}_{[O_1,r_1^2]}$$

are either located on the invariant center line O_1O_2, or they are the common points of the two circles $C[O_1, r_1]$ and $C[O_2, r_2]$ (if their intersection is non-empty).

Proof If the two circles intersect, then the common points are fixed, since either circle is point-wise invariant under the respected inversion. If the circles are tangent, then the common point is fixed and it also belongs to the center line O_1O_2.

So let us assume that the two circle have no point in common. Then, if a point of the plane $\mathcal{P}$, $A \in \mathcal{P}$ is mapped to A' by $\mathfrak{I}_{[O_1,r_1^2]}$ and then A' is mapped to A'' by $\mathfrak{I}_{[O_2,r_2^2]}$, then A is fixed by the composition of the inversions

$$\mathfrak{I}_{[O_2,r_2^2]} \circ \mathfrak{I}_{[O_1,r_1^2]}, \qquad \text{iff} \qquad A'' = A.$$

But then, O_1, A and A' are collinear and also O_2, A' and $A = A''$ are collinear. Also, $A' \neq A = A''$, since the two circles have no common point. Therefore, the four points O_1, $A = A''$, A', and O_2 are collinear, on the line O_1O_2.

Finally, the fixed point $A = A''$ is either a common point of the two circles, or is located on the line O_1O_2 and cannot belong to either circle (for otherwise, it would be a common point of the circles).

■

Step 2: If the two circles $C[O_1, r_1]$ and $C[O_2, r_2]$ do not intersect, then the Poncelet points are fixed under the composition of the inversions

$$\mathfrak{I}_{[O_2,r_2^2]} \circ \mathfrak{I}_{[O_1,r_1^2]}.$$

In **Figure 4.15**, the two given non-intersecting circles of inversion are the circles $C[O_1, R_1]$ and $C[O_2, R_2]$ and the circle $C[H, HP_1 = HP_2]$ is the Poncelet circle determined by them. Since the Poncelet circle $C[H, HP_1 = HP_2]$ is orthogonal to both $C[O_1, R_1]$ and $C[O_2, R_2]$, it is set-wise invariant under the above composition. If a point K is on the Poncelet circle, then $\mathfrak{I}_{[O_1,r_1^2]}$ maps it to a point N on the Poncelet circle and

$\mathfrak{I}_{[O_2,r_2^2]}$ maps it to L on the Poncelet circle, again. **Figure 4.15** shows the correspondence of points located on the Poncelet circle by the composition

$$\mathfrak{I}_{[O_2,r_2^2]} \circ \mathfrak{I}_{[O_1,r_1^2]}$$

(or its inverse $\mathfrak{I}_{[O_1,r_1^2]} \circ \mathfrak{I}_{[O_2,r_2^2]}$, if we move in reverse order).

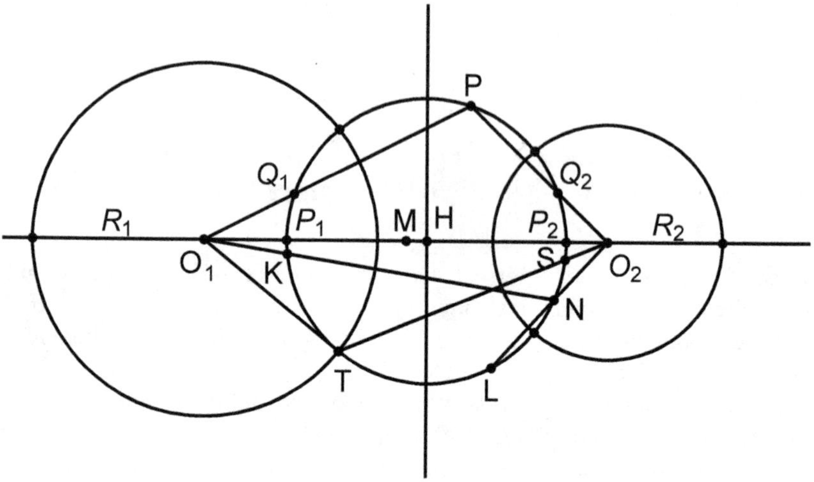

Figure 4.15: Correspondence of points on the Poncelet circle

Namely, we have depicted

$$Q_1 \longmapsto P \longmapsto Q_2, \qquad T \longmapsto T \longmapsto S, \qquad K \longmapsto N \longmapsto L.$$

We now see that *the Poncelet points are fixed by the composition*

$$I_{[O_2,c_2]} \circ I_{[O_1,c_1]} \qquad \text{(or its inverse} \quad I_{[O_1,c_1]} \circ I_{[O_2,c_2]}),$$

that is, we have, $P_1 \longmapsto P_2 \longmapsto P_1, \qquad P_2 \longmapsto P_1 \longmapsto P_2$. So, the Poncelet points are fixed under the composition $\mathfrak{I}_{[O_1,r_1^2]} \circ \mathfrak{I}_{[O_2,r_2^2]}$.

Step 3: We now find the equation that any fixed point of the composition $\mathfrak{I}_{[O_1,r_1^2]} \circ \mathfrak{I}_{[O_2,r_2^2]}$ must satisfy.

Suppose $\mathfrak{I}_{[O_1,r_1^2]}(A) = A'$, and $\mathfrak{I}_{[O_2,r_2^2]} = A''$. The point A on the line O_1O_2 is fixed, iff $A = A''$, that is, $\mathfrak{I}_{[O_2,c_2]} \circ \mathfrak{I}_{[O_1,c_1]}(A) = A'' = A$.

When this happens, we have $\overline{AO_2} = \overline{A''O_2}$, and so, with origin O_1 and axis $\overline{O_1O_2}$, we have that

$$\overline{O_1O_2} - \overline{O_1A} = \frac{-r_2^2}{\overline{O_2A'}} = \frac{-r_2^2}{\overline{O_1A'} - \overline{O_1O_2}} = \frac{-r_2^2}{\frac{r_1^1}{\overline{O_1A}} - \overline{O_1O_2}}.$$

Simplifying this, we find that in order for a point $A \in O_1O_2$ to be fixed, it must satisfy the **quadratic equation**, in $\overline{O_1A}$,

$$\overline{O_1O_2} \cdot \overline{O_1A}^2 - \left[r_1^2 - r_2^2 + (\overline{O_1O_2})^2\right] \cdot \overline{O_1A} + r_1^2\overline{O_1O_2} = 0.$$

Now, with $k = \overline{O_1O_2}$, $x = \overline{O_1A}$, origin O_1, and axis $\overline{O_1O_2}$, we obtain the quadratic equation

$$kx^2 - \left(k^2 + r_1^2 - r_2^2\right) x + kr_1^2 = 0.$$

Its two solutions are:

$$\overline{O_1A} = x_\pm = \frac{k^2 + (r_1 + r_2)(r_1 - r_2)}{2k} \pm$$
$$\pm \frac{\sqrt{(k + r_1 + r_2)(k + r_1 - r_2)(k - r_1 + r_2)(k - r_1 - r_2)}}{2k},$$

thus finding **equation 3.8**.

This is the equation for determining fixed points of the composition $\mathfrak{I}_{[O_2,c_2]} \circ \mathfrak{I}_{[O_1,c_1]}$ on the line O_1O_2. Depending on the parameters involved, we may have zero, one, or two such fixed points on the line O_1O_2, found by solving this quadratic equation. If the solutions are complex, then they correspond to two common points of the circle $C[O_1, r_1]$ and $C[O_2, r_2]$. If there is one double real solution, then the circles are tangent and this solution gives the common point of tangency. Finally if the solutions are two different real they give two fixed points on the center line O_1O_2, which must be the two Poncelet points, since we have previously shown that the Poncelet points are fixed under the above composition.

In conclusion, the Poncelet points P_1 and P_2 are the only two points on the line O_1O_2 which are fixed by $\mathfrak{I}_{[O_2,r_2^2]} \circ \mathfrak{I}_{[O_1,r_1^2]}$ (and / or $\mathfrak{I}_{[O_1,r_1^2]} \circ \mathfrak{I}_{[O_2,r_2^2]}$). They satisfy the above quadratic equation of the fixed points on the center line O_1O_2. In this case, we have $O_1O_2 > r_1 + r_2$, or $0 < O_1O_2 < |r_1 - r_2|$, and so the quadratic equation has two real solutions, which represent the Poncelet points. By solving this quadratic equation, we find the positions of P_1 and P_2 on the axis $\overline{O_1O_2}$ with origin O_1, in terms of r_1, r_2, and $\overline{O_1O_2}$.

(If $O_1O_2 = r_1 + r_2$, or $0 < O_1O_2 = |r_1 - r_2|$, then the circles are tangent externally, or internally, and the equation has one real double solution that locates the point of tangency of the circles. This is the one double point of Poncelet. If $O_1O_2 < r_1 + r_2$, then the circles intersect at two points, and the equation has two complex conjugate solutions whose real and imaginary parts represent the coordinates of these two points, if the positive real axis is the straight line $\overline{O_1O_2}$ with origin O_1. So, in this case, we can say that the Poncelet points are the points of intersection of the two circles.)

Inversion Preserves Ratios Similar to Cross-Ratio.

One way to prove the preservation of the value of a ratio similar to cross-ratio, is to use the **Distance formula for Inverse Points**. Apart from that, this formula is very useful in many occasions.

We consider an inversion $\mathfrak{I}$ with center O and power $c \neq 0$. For any two points A and B in the plane of the inversion, we let A' and B' be their inverse points, respectively. Then,

$$\overline{OA} \cdot \overline{OA'} = \overline{OB} \cdot \overline{OB'} = c,$$

and so,

$$OA \cdot OA' = OB \cdot OB' = |c|.$$

We would like to find the distance of A' and B' by means of the inversion.

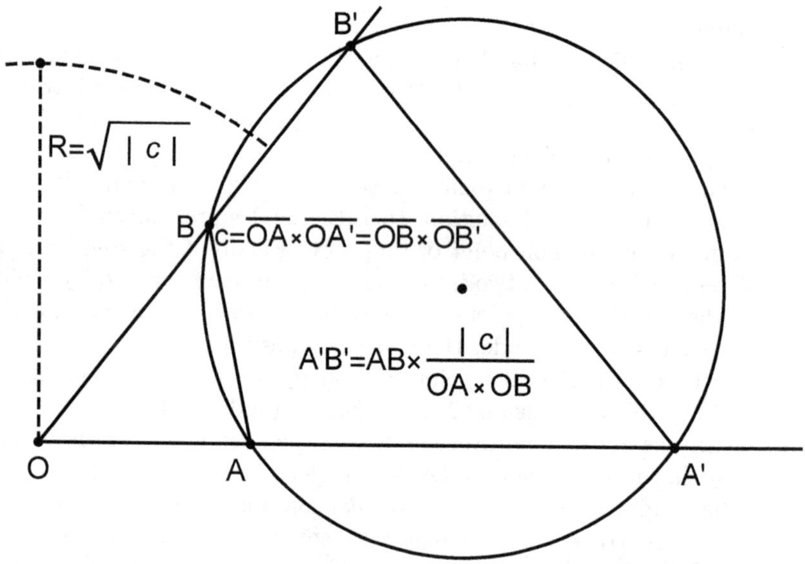

Figure 4.16: For the distance between inverse points (circle case)

(I) Assume first that the points A, B, A' and B' are **concyclic**. See **Figure 4.16**. Then, the triangles OAB and $OA'B'$ are (oppositely) similar. Hence, we have

$$\frac{OA}{OB'} = \frac{OB}{OA'} = \frac{AB}{A'B'}.$$

So,

$$A'B' = AB \cdot \frac{OA'}{OB} = AB \cdot \frac{OA' \cdot OA}{OB \cdot OA} = AB \cdot \frac{|c|}{OA \cdot OB}.$$

Finally, **the distance formula for the inverse points** is

$$A'B' = AB \cdot \frac{|c|}{OB \cdot OA}.$$

Remark: If OP and OP' is the distance of O from the lines AB and $A'B'$, respectively, the heights of the triangle OAB and $OA'B'$. Then, the similarity of the triangles OAB and $OA'B'$ proves the following proportionality for these heights

$$\frac{A'B'}{AB} = \frac{OP'}{OP} \qquad \text{or} \qquad \frac{A'B'}{OP'} = \frac{AB}{OP}.$$

(II) Next, we assume that the points A, B, A' and B' are ***collinear***, which is the case if O, A, and B are collinear. See **Figure 4.17**.

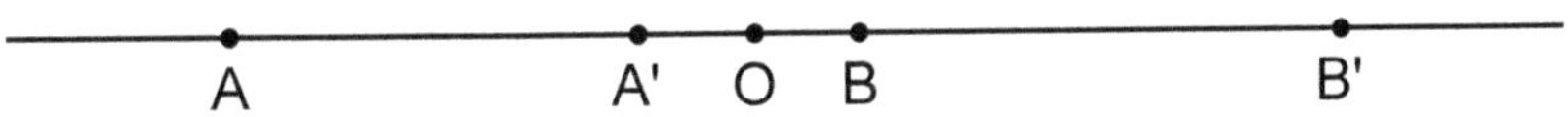

Figure 4.17: For the distance between inverse points (straight line case)

Then,

$$\overline{A'B'} = \overline{A'O} + \overline{OB'} = -\frac{c}{\overline{OA}} + \frac{c}{\overline{OB}} = \frac{c(\overline{OA} - \overline{OB})}{\overline{OB} \cdot \overline{OA}} = \frac{c \cdot \overline{BA}}{\overline{OB} \cdot \overline{OA}} = \frac{-c \cdot \overline{AB}}{\overline{OB} \cdot \overline{OA}}.$$

So, **the distance formula for the inverse points** is

$$A'B' = \frac{|c| \cdot AB}{OB \cdot OA},$$

which is a formula similar to the formula of case **(I)**.

Next, we consider four points A, B, C and D in the plane and their inverses A', B', C' and D' under an inversion $\mathfrak{J}$ with center O and power $c \neq 0$. Then, it holds

$$\frac{AC}{AD} \div \frac{BC}{BD} = \frac{AC \cdot BD}{AD \cdot BC} = \frac{A'C'}{A'D'} \div \frac{B'C'}{B'D'} = \frac{A'C' \cdot B'D'}{A'D' \cdot B'C'}.$$

This follows from the distance formula of inverse points found above. Namely, by applying the distance formula and simplifying, we get

$$\frac{A'C'}{A'D'} = \frac{AC}{AD} \cdot \frac{OD}{OC} \qquad \text{and} \qquad \frac{B'D'}{B'C'} = \frac{BD}{BC} \cdot \frac{OC}{OD}.$$

By multiplying these two equations side-wise, we obtain the claimed equality.

If the eight points are on the same straight line, considered as an axis going through O necessarily, we find the same relation with oriented segments. Namely:

$$\frac{\overline{AC}}{\overline{AD}} \div \frac{\overline{BC}}{\overline{BD}} = \frac{\overline{AC} \cdot \overline{BD}}{\overline{AD} \cdot \overline{BC}} = \frac{\overline{A'C'}}{\overline{A'D'}} \div \frac{\overline{B'C'}}{\overline{B'D'}} = \frac{\overline{A'C'} \cdot \overline{B'D'}}{\overline{A'D'} \cdot \overline{B'C'}}. \tag{4.3}$$

Apropos of this equation, we give the following **standard definition:**

Given four points A, B, C and D on a straight line or better an axis, the quantity

$$\{A, B; C, D\} := \frac{\overline{AC}}{\overline{AD}} \div \frac{\overline{BC}}{\overline{BD}} = \frac{\overline{AC} \cdot \overline{BD}}{\overline{AD} \cdot \overline{BC}} = \frac{\overline{AC}}{\overline{BC}} \div \frac{\overline{AD}}{\overline{BD}} = \frac{\overline{CA}}{\overline{CB}} \div \frac{\overline{DA}}{\overline{DB}}, \quad \text{etc.}$$

is called the ***cross-ratio (or double ratio) of*** *these* ***four points.***[1]

So, we have proved, above, that ***the cross-ratio is preserved under inversion.***

Furthermore, we see that the cross-ratio can be written in several ways. The cross-ratio has many properties and applications in several branches of mathematics. (See many properties of the cross ratio in the exercises.) The property of preservation of the cross-ratio by inversion is very important and useful. For example, see the complete **proof of Steiner's Porism**, given also in this chapter.

Also, in geometry, **a quadruple of points** $\{A,\ B;\ C,\ D\}$ **on an axis is harmonic** iff

$$\{A,\ B\,;\ C,\ D\} := -1 \qquad \Longleftrightarrow \qquad -\frac{\overline{CA}}{\overline{CB}} = \frac{\overline{DA}}{\overline{DB}}, \qquad \text{etc.}$$

Here, C must be between A and B and B between C and D. (If the order of the four points is such, then their cross ratio is negative. But, if both C and D are between A and C, then their cross ratio is positive, etc.) Hence, by what we have proved, if **the quadruple of points** $\{A,\ B\,;\ C,\ D\}$ **on an axis is harmonic**, then $\{A',\ B'\,;\ C',\ D'\} := -1$, and so, the quadruple of the inverse points $\{A',\ B'\,;\ C',\ D'\}$ is also harmonic (and vice-versa).

[1]Other authors defined the **cross-ratio (or double ratio)** in different equivalent ways. Read carefully the exact definition in the book that you study and check its equivalence with the definition we cite here.

About Angle Preservation by Inversions.

The angle of two curves at a point of their intersection is the angle between their tangent lines at that point, if of course each of these tangents exists.

The tangent line to a curve at a point A is the limit position of the secant lines AB, as B moves on the curve and approaches A. See **Figure 4.18**. If this limit exists, then the tangent exists. Otherwise the tangent does not exist. If a tangent exists, then it is unique, because the limit position is unique.

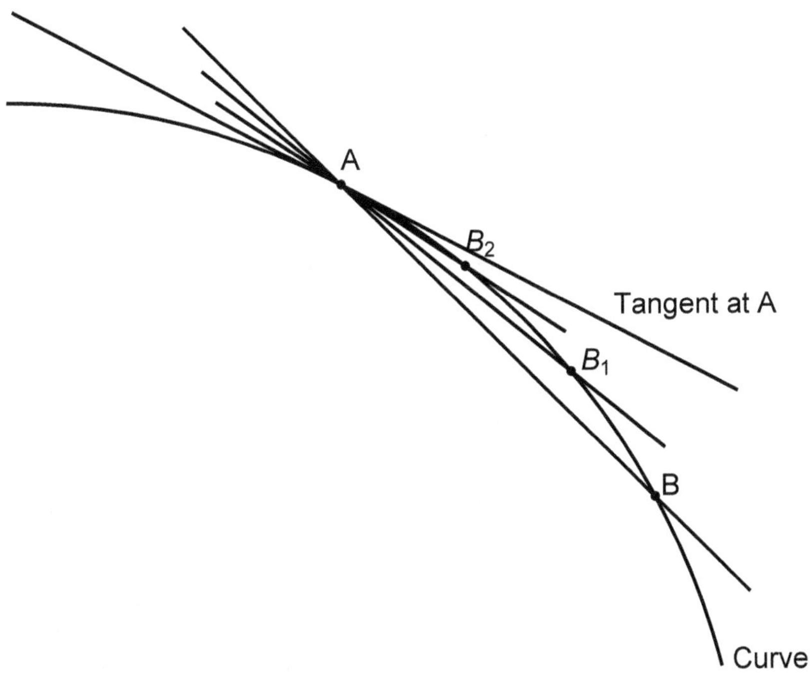

Figure 4.18: Tangent to a curve at a point A is the limit position of the secant lines AB, as B approaches A.

According to this definition a circle has tangent at any of its points, which is the usual geometrical tangent perpendicular to the radius at the point of tangency. A straight line has tangent at any point of it, which is the straight line itself. Also, this definition is the same as the definition of tangent in a course of calculus.

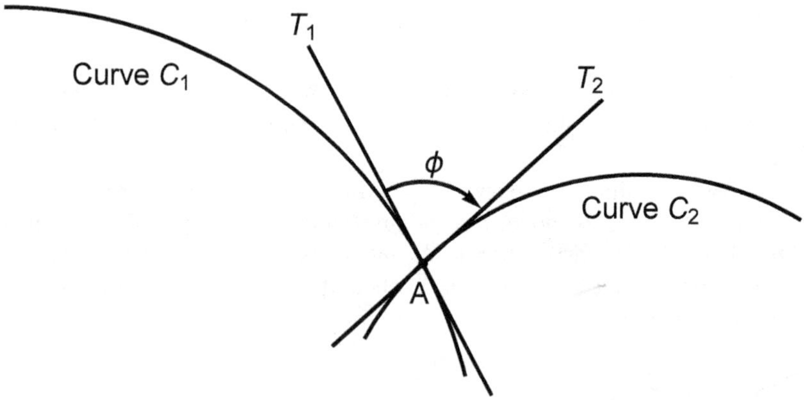

Figure 4.19: Oriented angle from curve C_1 to curve C_2

So, if two curves C_1 and C_2 intersect at a point A, and both have tangent AT_1 and AT_2, respectively, at A, then the angle between the two curves is the angle between their tangents. See **Figure 4.19**. In geometry, this angle may be considered non-oriented and its measure be a positive number. But, in more precise situations, we need to consider this angle to be oriented. Then, we write

$$\phi = \measuredangle(C_1, C_2) = \measuredangle(AT_1, AT_2),$$

opposite to

$$-\phi = \measuredangle(C_2, C_1) = \measuredangle(AT_2, AT_1).$$

We have seen that similarities preserve the absolute value of an angle, but the orientation is preserved by the direct similarities and is reversed by the opposite similarities. We will show that inversions preserve the absolute value of an angle and reverse the orientation. It is enough to prove this with positive inversions since any negative inversion is the commutative composition of a positive inversion and a half-turn, and the half-turns preserve oriented angles. So, in **the following Lemma and Theorem**, without loss of generality, we consider the inversion to be positive.

Lemma *An inversion reverses the oriented angles between a ray through the center of inversion and two inverse curves at the corresponding inverse points of intersection.*

Proof We consider an inversion with center O and two inverse curves C_1 and C_2. We draw any ray Ox and keep it fixed. Let Ox intersect the curves C_1 and C_2 at P and P', respectively, and let PT and $P'T'$ be the corresponding tangents to the curves C_1 and C_2 at these points. See **Figure 4.20**.

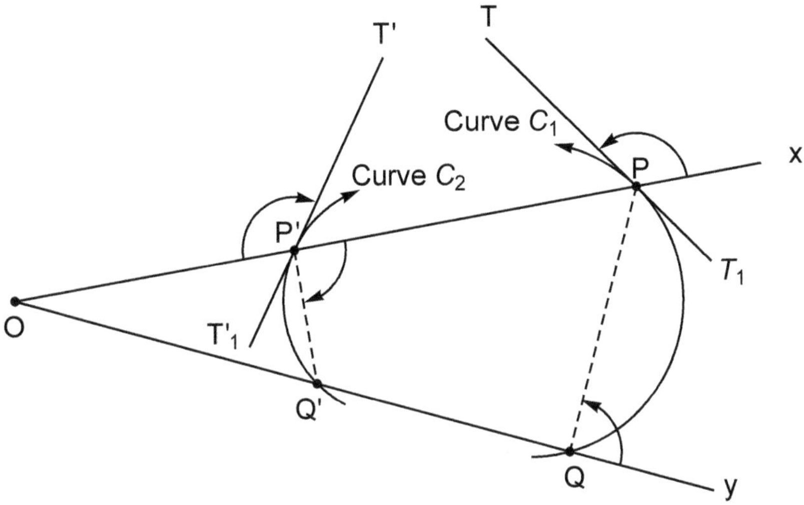

Figure 4.20: Angles from a ray through the center of inversion to two inverse curves

Now, we draw any other ray Oy that intersects the curves C_1 and C_2 at Q and Q', respectively. Since

$$\overline{OP} \cdot \overline{OP'} = \overline{OQ} \cdot \overline{OQ'} (= c),$$

the four points P, P', Q, and Q' are concyclic (including the straight line as an infinite circle). So, the angles $\measuredangle(Qy, QP)$ and $\measuredangle(P'x, P'Q')$ are opposite.

Next, we let Q approach P. Then, the ray Oy approaches the ray Ox, point Q' approaches P' and the lines PQ and $Q'P'$ approach the tangents PT and $P'T'$.

Hence,

$$\lim_{Q \to P} \measuredangle(Qy, QP) = \measuredangle(Px, PT)$$

and

$$\measuredangle(Qy, QP) = -\measuredangle(P'x, P'Q').$$

So,

$$\lim_{Q \to P} \measuredangle(P'x, P'Q') \;=\; \measuredangle(P'x, P'T_1') \;=\; \measuredangle(P'O, P'T').$$

Thus we obtain,

$$\measuredangle(Px, PT) = -\measuredangle(P'O, P'T').$$

■

Remark: We observe that the tangents PT and $P'T'$ are either parallel or if they intersect at a point D (not put in the **Figure**), then the triangle PDP' is isosceles. Therefore D is on the perpendicular bisector of the segment PP'. If one of the (two different) curves is a straight line (and so the other curve is a circle, and vice-versa), then D is on this straight line.

Now, we can prove the following important Theorem:

Theorem 4.4.4 *An inversion reverses oriented angles. That is, given two curves that intersect at a point P, their inverse curves intersect at P', the inverse of P, and the oriented angles of the original curves at P are opposite to the corresponding oriented angles of their inverse curves at P'. So, an inversion preserves the absolute values of the oriented angles and reverses their orientation.*

Proof We consider two curves C_1 and C_2 that intersect at P and their corresponding tangents PT_1 and PT_2, respectively. We also consider their inverse curves C_1' and C_2' that intersect at the point P', inverse of P and their corresponding tangents $P'T_1'$ and $P'T_2'$, respectively. See **Figure 4.21**.

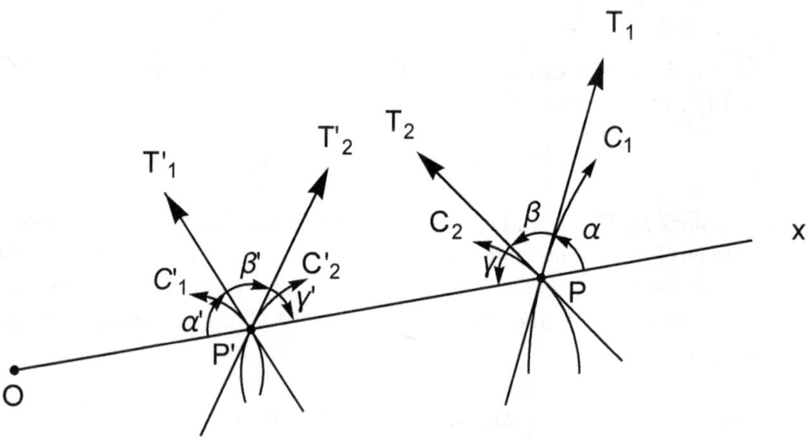

Figure 4.21: An inversion reverses the oriented angle between two curves and their inverses

We draw the ray Ox through the center of inversion O and containing the inverse points P and P'. We let

$$\alpha = \measuredangle(Px, PT_1), \quad \beta = \measuredangle(PT_1, PT_2), \quad \gamma = \measuredangle(PT_2, PO),$$

and

$$\alpha' = \measuredangle(P'O, P'T_1'), \quad \beta' = \measuredangle(P'T_1', P'T_2), \quad \gamma' = \measuredangle(P'T_2, P'x),$$

and we have:

$$\alpha + \beta + \gamma = \pi, \quad \text{and} \quad \alpha' + \beta' + \gamma' = -\pi,$$
$$\alpha = -\alpha', \quad \text{and} \quad \gamma = -\gamma'.$$

(The last two equalities are obtained by the **previous Lemma.**)

These four equalities imply $\beta = -\beta'$. That is,

$$\measuredangle(C_1, C_2) = \measuredangle(PT_1, PT_2) = -\measuredangle(P'T_1', P'T_2') = -\measuredangle(C_1', C_2').$$

■

When the Composition of two Positive Inversions is Commutative.

Theorem 4.4.5 *We assume that the centers of two inversions are different, $O_1 \neq O_2$, and both powers are positive, $c_1 > 0$ and $c_2 > 0$, then:*

$$I_{[O_2,c_2]} \circ I_{[O_1,c_1]} = I_{[O_1,c_1]} \circ I_{[O_2,c_2]}$$

(this composition commutes) if and only if the circles of the two inversions are orthogonal [so, $(O_1O_2)^2 = c_1 + c_2 = R_1^2 + R_2^2$].

Proof. ($\Longrightarrow$) We assume that the circles of the two inversions with centers O_1 and O_2 are orthogonal and intersect at A and B and P a point in their plane. We let $I_{[O_1,c_1]}(P) = P_1$, $I_{[O_2,c_2]}(P) = P_2$, $I_{[O_1,c_1]}(P_2) = X$, and $I_{[O_2,c_2]}(P_1) = Y$. See **Figure 4.22**.

We have the inverse pair of points P and P_1 under $I_{[O_1,c_1]}$. The inverses of these points under the inversion $I_{[O_2,c_2]}$, are P_2 and say X, respectively. Also, the inverse of the circle $C[O_1, c_1]$ under the second inversion is itself, since the two circles $C[O_1, c_1]$ and $C[O_2, c_2]$ are orthogonal. Therefore, P_2 and X are inverses with respect to the circle $C[O_2, c_2]$. That is, $X = Y$ and so, *we also obtain that P_1, X, P_2 and P are concyclic,* as (P, P_1) and (P_2, X) are pairs of inverse points in the circle $C[O_1, c_1]$.

($\Longleftarrow$) Now, we assume that the composition commutes and we want to prove that the circles of the two inversions are orthogonal. We use

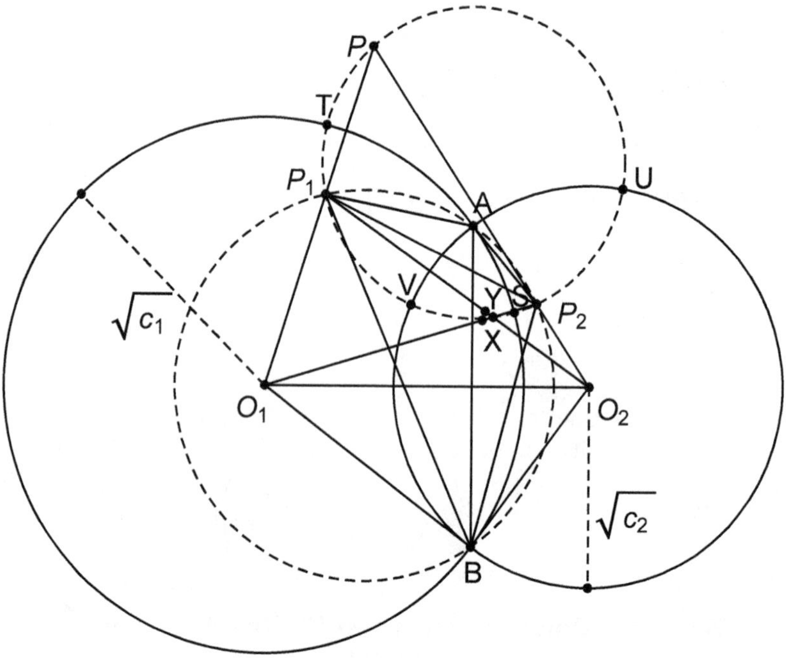

Figure 4.22: The circles of the two inversions are orthogonal if and only if $I_{[O_2,c_2]} \circ I_{[O_1,c_1]} = I_{[O_1,c_1]} \circ I_{[O_2,c_2]}$

the contrapositive argument. So we suppose that the circles of the two inversions are not orthogonal. See **Figure 4.23**.

Then, the line O_1A is not tangent to the circle $C[O_2, R_2]$ and so it intersects this circle at a second point C, which may be outside or inside the circle $C[O_1, R_1]$. Assume that C is outside $C[O_1, R_1]$. {We do analogous work if C is inside $C[O_1, R_1]$. Also notice that $I_{[O_2,c_2]}(C) = C$.}

Let $D := I_{[O_1,c_1]}(C)$ and $E := I_{[O_2,c_2]}(D)$. Then D is inside $C[O_1, R_1]$ and outside $C[O_2, R_2]$ and so E is inside $C[O_2, R_2]$. Therefore,
$D \neq A$, $E \neq D$, and $I_{[O_2,c_2]} \circ I_{[O_1,c_1]}(C) = E$ and
$I_{[O_1,c_1]} \circ I_{[O_2,c_2]}(C) = I_{[O_1,c_1]}(C) = D$.

Since $D \neq E$, we have that $I_{[O_2,c_2]} \circ I_{[O_1,c_1]} \neq I_{[O_1,c_1]} \circ I_{[O_2,c_2]}$.

■

Remark: If the two inversions have the same center $O_1 = O_2 := O$, then the composition $I_{[O,c_2]} \circ I_{[O,c_1]}$ is a homothety with center O and homothetic ratio $\frac{c_2}{c_1}$. (The proof is straightforward. So, this composition is

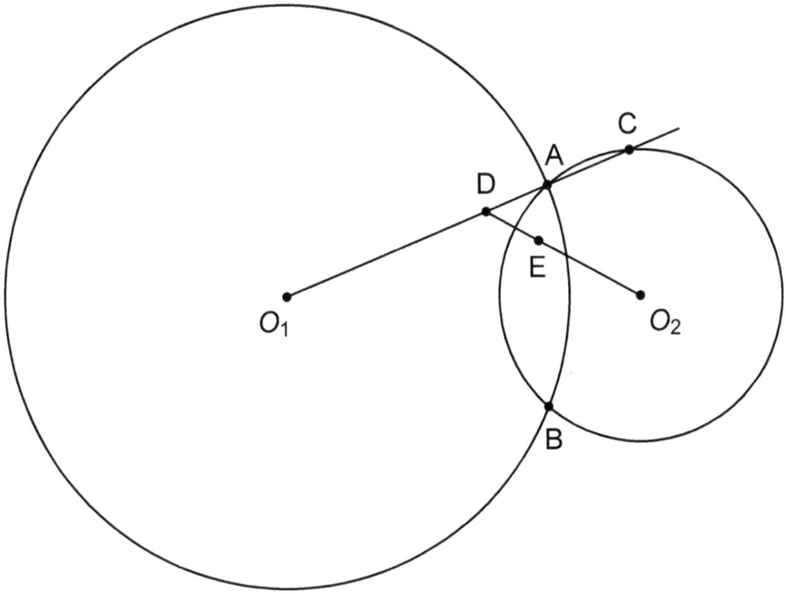

Figure 4.23: The circles of the two inversions are not orthogonal. Then, $I_{[O_2,c_2]} \circ I_{[O_1,c_1]} \neq I_{[O_1,c_1]} \circ I_{[O_2,c_2]}$

not commutative, unless $c_2 = c_1$, in which case the homothety is identity, or $c_2 = -c_1$, in which case the homothety is a half turn.)

Invariance of Inversiveness under Inversion.

Theorem 4.4.6 (Invariance of inversiveness under inversion.) *Suppose two given points P and P_1 are inverse of each other with respect to a given circle $C[Q, r]$ and the whole figure of $C[Q, r]$ and P and P_1 is inverted with respect to another circle $C[O, \sqrt{|c|}]$. Then, in the resulting figure, the images P' and P_1' of the given points P and P_1 are inverse with respect to $I_{[O,c]}\{C[Q, r]\}$, the image of the given circle $C[Q, r]$ under the inversion $I_{[O,c]}$.*

We must clarify that **in this Theorem** straight lines are considered circles with center a point at infinity. We know that lines and circle may be inverted to other lines and or circles interchangeably.

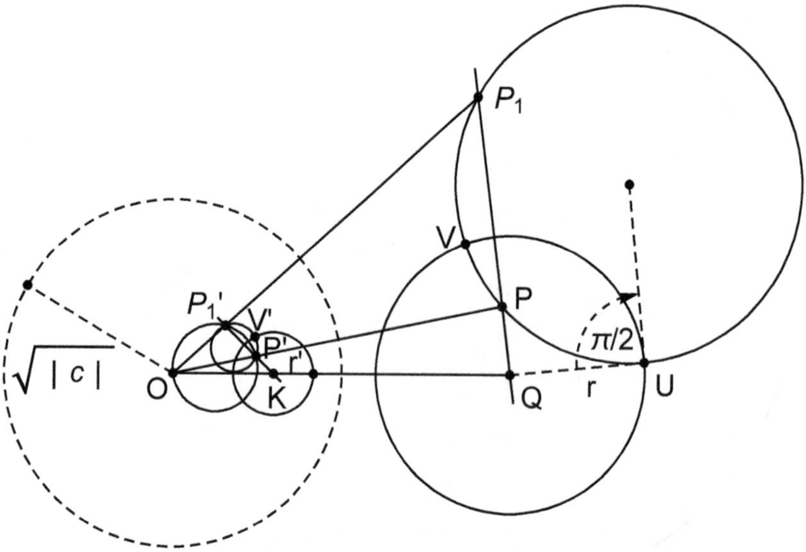

Figure 4.24: Invariance of inversiveness under inversion

Proof (1) Suppose the given circle is $C[Q,r]$ and two inverse points with respect to it are P and P_1, as in **Figure 4.24**.

We invert this circle and the two inverse points with respect to another circle $C[O,\sqrt{|c|}]$, for which $O \notin C[Q,r]$. The circle $C[Q,r]$ is mapped to the circle $C[K,r']$ and the points P and P_1 to P' and P_1', respectively. We must prove that P' and P_1' are inverses with regard to the new circle $C[K,r']$.

We draw either the radial line QPP_1 or any two circles through the points P and P_1. The line or these circles are orthogonal to $C[Q,r]$. Since inversion preserves orthogonality the inverses of the radial line QPP_1 and the circles through the points P and P_1 are either circles or lines orthogonal to the circle $C[K,r']$ and pass through P' and P_1'. In **Figure 4.24**, we have inverted the radial line QPP_1 to the circle $(OP'P_1')$ and the circle (PUP_1V) to the circle $(P'P_1'V')$. These new circles intersect the circle $C[K,r']$ orthogonally.

Therefore, the points P' and P_1' are inverse points with respect to $C[K,r']$. (Thus, the points K, P', and P_1' are collinear.)

(2) In **Figure 4.25**, the center O of the new circle $C[O,\sqrt{|c|}]$ is on the circle $C[Q,r]$. So, by the new inversion $I_{[O,c]}$, the circle $C[Q,r]$ becomes

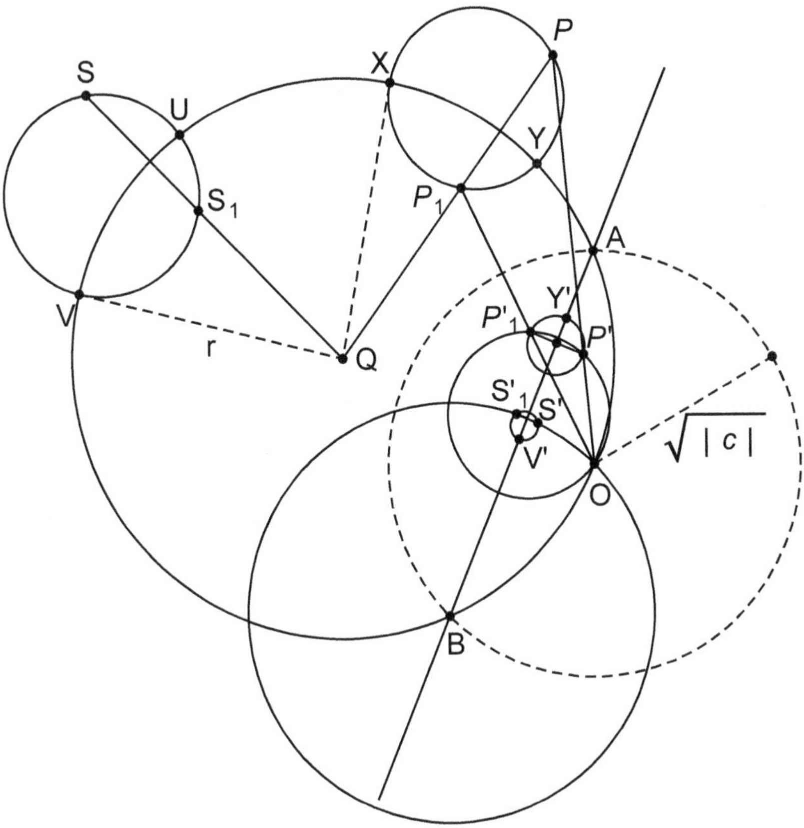

Figure 4.25: Invariance of inversiveness under inversion

the straight line AB through the common points A and B of $C[Q, r]$ and $C[O, \sqrt{|c|}]$.

The images (P', P_1') and (S', S_1'), under the inversion $I_{[O,c]}$, with respect to $C[O, \sqrt{|c|}]$, of two pairs of inverse points (P, P_1) and (S, S_1) with regard to the circle $C[Q, r]$, are now symmetrical with respect to the line AB, as it is seen in **Figure 6.56**, since the inverses of the radial lines QP_1P and QS_1S, and of the circles (PXP_1Y) and (SUS_1V), are new circles (or straight lines) orthogonal to the line AB, and so the line AB is a diameter to the new circles.

■

The following corollary is very important and supplements **Theorem 4.4.2**.

Corollary 4.4.1 *Suppose two figures $\mathfrak{F}_1$ and $\mathfrak{F}_2$ are mutually inverse under a positive inversion $I_{[O,c]}$, $(c > 0)$. The inverses of $\mathfrak{F}_1$ and $\mathfrak{F}_2$ under any inversion $I_{[Q,k]}$ with $Q \in C[O, \sqrt{c}]$ any point of the circle of the first inversion, are symmetrical in a straight line and therefore isometric. In particular, two inverse circles invert under $I_{[Q,k]}$ into equal circles.*

Proof Since $Q \in C[O, \sqrt{c}]$, the circle $C[O, \sqrt{c}]$ inverts into a straight line under $I_{[Q,k]}$. Then, by the **previous Theorem** the figures $\mathfrak{F}_1$ and $\mathfrak{F}_2$ invert into figures which are mutually inverse with respect to this straight line. Therefore, they are symmetrical in this line.

■

Steiner's Porism and its Proof.

Theorem 4.4.7 (Steiner's Porism) *For Steiner chains within nested base circles we have:*
If one chain of circles between two given nested circles closes (that is, it becomes periodic), then we have the following results:
(a) Every chain between the two given circles closes.
(b) So, there are infinitely many closed chains between the two given circles.
(c) Any circle tangent to the two given nested circles belongs to a closed chain.
(d) The number of circles in every closed chain is the same.
(e) Therefore: If one chain does not close, then no chain closes.

Proof Since the two base nested circles do not intersect, as we have seen in **Theorem 4.3.1**, they can be inverted to concentric circles. Then, any chain between them is inverted to a chain of image-circles between these two concentric circles, by the same inversion. So, the circles of such a chain invert to equal circles, tangent to the two new base concentric circles and vice-versa.

So, if the Porism is valid for concentric base circles (they form a **canonical annulus**), then applying the same inversion to the canonical annulus and its contained circles, we transfer the result back to the original annulus.

In fact, Steiner proved that an appropriate condition between the radii and center-segment and a positive integer n exists, so that a chain closes after n circles of the chain have been constructed. It is enough to solve the problem in the canonical annulus of the two image-base circles. Then, no matter where we start in the original annulus, the chain will always close after n circles have been created, because the characteristics of the figure involving the canonical annulus are rotation invariant about its center. See **Figure 4.26**, next.

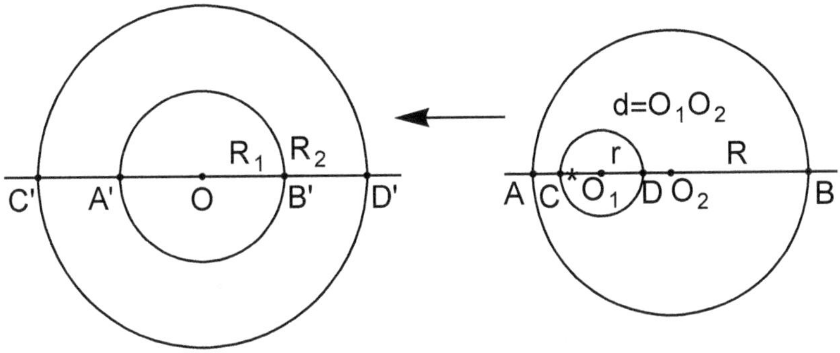

Figure 4.26: Conditions for non-intersecting circles to invert into concentric circles that admit a closing chain

Now, to complete the proof of the Porism, we must prove the existence of closing ordinary chains. We will derive **Steiner's formula** or criterion that the two base circles must satisfy in order for any chain inside their annulus to close.

In **Figure 4.26**, we have two non-concentric circles with centers O_1 and O_2 and diameters CD and AB, respectively. The radii are $O_1C = O_1D = r$, $O_2A = O_2B = R$ and the center-segment $O_1O_2 = d$.

We consider the inversion claim by **Theorem 4.3.1**, which maps the circles $C[O_1, r]$ and $C[O_2, R]$ onto concentric circles with common center O and corresponding diameters $C'D'$ and $A'B'$. The radii of the new circles are $OC' = OD' := R_1$ and $OA' = OB' := R_2$, respectively.

Since inversions preserve the cross-ratio of four points on an axis, as we have seen in **equation 4.3**,we obtain

$$[A, B; C, D] = [A', B'; C', D'], \qquad \text{or,} \qquad \frac{\overline{AC} \cdot \overline{BD}}{\overline{AD} \cdot \overline{BC}} = \frac{\overline{A'C'} \cdot \overline{B'D'}}{\overline{A'D'} \cdot \overline{B'C'}}.$$

We work out either side of this equality and we find that: the *right-hand side* of this equality equals

$$\frac{(R_1 - R_2)^2}{(R_1 + R_2)^2}$$

and the *left-hand side* equals

$$\frac{(R - d - r)(R + d - r)}{(R - d + r)(R + d + r)} = \frac{(R - r)^2 - d^2}{(R + r)^2 - d^2}.$$

We equate these two results and simplify to obtain the relation

$$(R-r)^2 - d^2 = Rr\left(\frac{R_1}{R_2} + \frac{R_2}{R_1} - 2\right). \tag{4.4}$$

(Use the properties of proportions, or work it out in your way.)

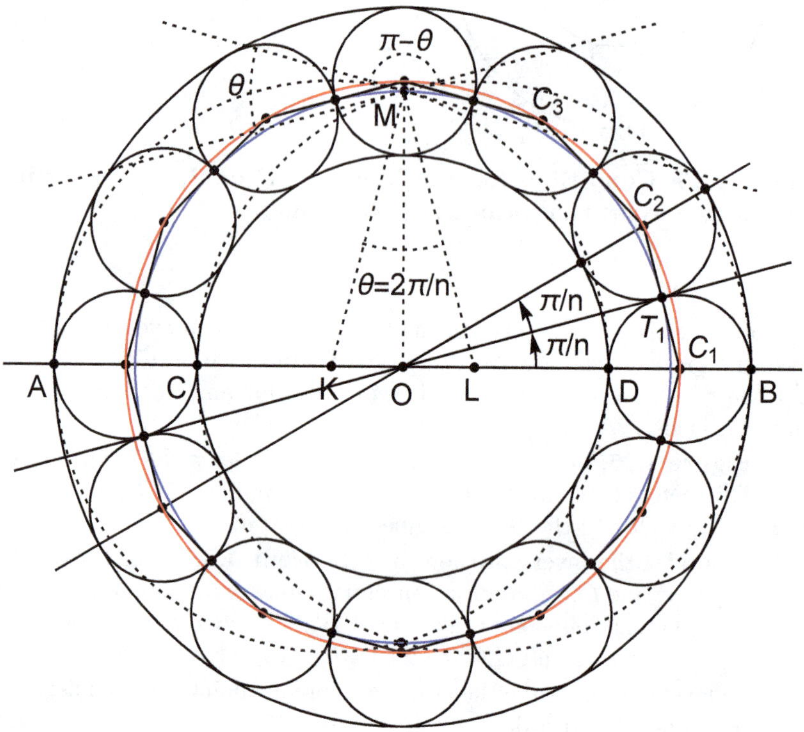

Figure 4.27: Closing simple canonical Steiner chain of circles between two concentric base circles.

Now in **Figure 4.27**, we have a canonical annulus (annulus between two concentric circles) and a closing chain of n equal circles. Then, the radius of the equal circles of this chain is

$$C_1T_1 = C_2T_1 = \frac{R_2 - R_1}{2}.$$

We also have,

$$OC_1 = OC_2 = \frac{R_2 + R_1}{2}$$

and

$$OT_1 = \sqrt{OC_1^2 - C_1T_1^2} = \sqrt{R_1R_2}\,.$$

So, from the right triangle T_1OC_1, we get

$$\sin\left(\frac{\pi}{n}\right) = \frac{R_2 - R_1}{R_2 + R_1} \iff \frac{R_2}{R_1} = \frac{1+\sin\left(\frac{\pi}{n}\right)}{1-\sin\left(\frac{\pi}{n}\right)}.$$

We substitute this ratio into the **equation 4.4** above and simplify to obtain

$$(R-r)^2 - d^2 = 4Rr\tan^2\left(\frac{\pi}{n}\right) \iff (R+r)^2 - d^2 = 4Rr\sec^2\left(\frac{\pi}{n}\right),$$

and so,

$$d = \sqrt{(R-r)^2 - 4Rr\tan^2\left(\frac{\pi}{n}\right)} \geq 0 \iff$$
$$d = \sqrt{(R+r)^2 - 4Rr\sec^2\left(\frac{\pi}{n}\right)} \geq 0.$$

This is **Steiner's formula** (written in four forms) that R, r, d, and n must satisfy for a chain inside the original annulus to close after n circles have been placed so that, any two successive circles are tangent and all the circles are tangent to the base circles.

When the base circles are concentric, $d = 0$. Then, we can solve for R, the radius of the outer circle, or for r, the radius of the inner circle, in terms of the number n and r or R, respectively, in order to construct a canonical closed chain of n circles. Notice that in this case, the radius of each circle of this canonical chain is $\rho := \dfrac{R-r}{2}$. So we have:

$$\tan^2\left(\frac{\pi}{n}\right) = \frac{(R-r)^2}{4Rr}\left(= \frac{\rho^2}{Rr}\right).$$

This implies,

$$\sin^2\left(\frac{\pi}{n}\right) = \frac{(R-r)^2}{(R+r)^2}, \quad \text{and so,} \quad \sin\left(\frac{\pi}{n}\right) = \pm\frac{R-r}{R+r}.$$

But for $n = 2,\ 3,\ 4\ldots,\ 0 < \dfrac{\pi}{n} \leq \dfrac{\pi}{2}$, and so $\sin\left(\dfrac{\pi}{n}\right) > 0$. Then, when $d = 0$ we have,

$$R > r \implies R = r\cdot\frac{1+\sin\left(\frac{\pi}{n}\right)}{1-\sin\left(\frac{\pi}{n}\right)}, \quad \text{and} \quad R \leq r \implies R = r\cdot\frac{1-\sin\left(\frac{\pi}{n}\right)}{1+\sin\left(\frac{\pi}{n}\right)}.$$

Now, all the claims, stated in the Porism, follow easily from the results obtained in this proof. This essentially finishes the proof of Steiner's Porism.

■

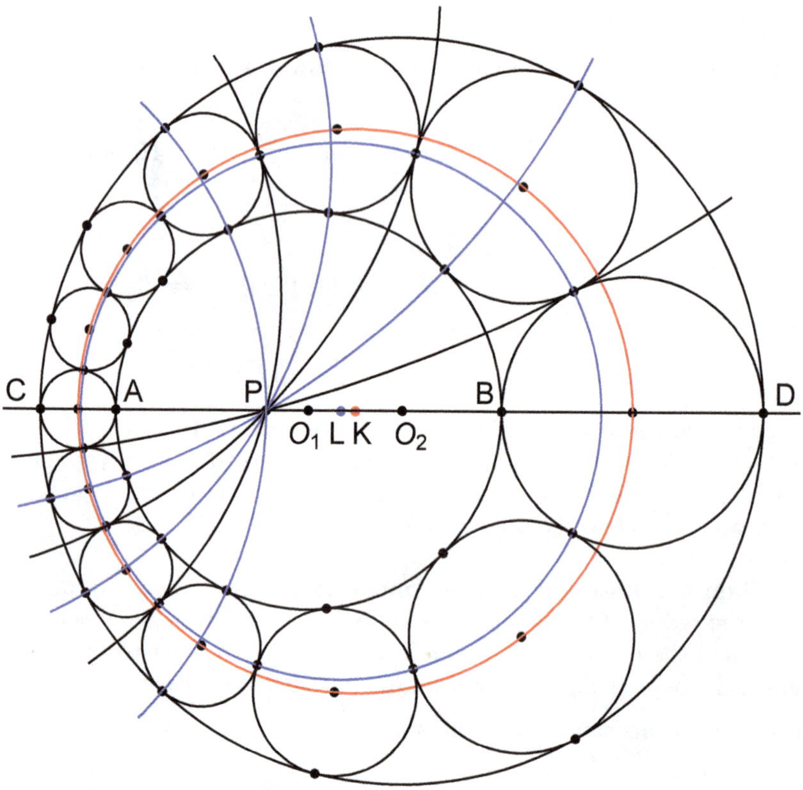

Figure 4.28: Closing simple Steiner chain of even number circles between two non-concentric circles

Remarks: (a) Notice that R is obtained by both $\alpha := \dfrac{\pi}{n}$ and $\pi - \alpha$. Check these formulae for R with $n = 1,\ 2,\ 3,\ 4,\ 5$ and 6, and interpret the results.

(b) For $n = \infty$ and $R > r$, we get $d = |R - r| = R - r$. Then, the circles are internally tangent and the chain is a Pappus chain.

Let us now illustrate some results of the **Steiner's Porism** and its proof. **Figure 4.28** is the inverse of **Figure 4.27**, under an inversion with center a point P, thus producing a Steiner chain between two non-concentric circles. The point P is one of two Poncelet points of $C[O_1, r]$ and $C[O_2, R]$, the inverses of the two given base circles, with

$$R = O_2C = O_2D > r = O_1A = O_1B$$

and center-segment $d = O_1O_2$. The radical axis of the circles $C[O_1, r]$ and $C[O_2, R]$, as they are situated in the figure, is located outside of them and to the left of P.

We must observe that the lines, such as OT_1, etc., in **Figure 4.27**, are inverted to circles orthogonal to the two base boundary circles, passing through the center of inversion, P, and tangent to the circles of the chain at the corresponding points. Similarly, the lines as OC_2 are inverted to circles orthogonal to both the base circles and the circles of the chain at the corresponding points, just as the line OC_1 is inverted to the line AD which passes through the center of inversion P. All the circles through P and orthogonal to the two base circles also intersect at the second Poncelet point of the two base circles. (In the figure, these circles are not drawn fully and so the second Poncelet point is not shown.)

4.5 A Special Composition of Inversions

Composition of Inversions of the Type $I_{[O,R_1^2]} \circ I_{[Q,R_2^2]} \circ I_{[O,R_1^2]}$.

The composition of inversions is not a new inversion, except for special cases. In general, for the composition of n inversions $F = I_1 \circ I_2 \circ \ldots \circ I_n$ we can say:

(a) F maps circles to circles or straight lines, and straight lines to circles or straight lines.

(b) If n is even, then F preserves the oriented angles.

(c) If n is odd, then F reverses the oriented angles but preserves their absolute values.

(d) Given two positive inversions $I_1 = I_{[O,R_1^2]}$ and $I_2 = I_{[Q,R_2^2]}$, we have that, $I_1 \circ I_2 = I_2 \circ I_1$ (i.e., their composition is commutative), if and only if, the circles $\mathcal{C}(O, R_1)$ and $\mathcal{C}(Q, R_2)$ of I_1 and I_2 are orthogonal (iff $OQ^2 = R_1^2 + R_2^2$). (See, Roussos, 2022, Theorem 6.5.6, pp. 368, or **Chapter 4, Theorem 4.4.5**.) However, this composition is not an inversion, even though it is involutary (why?).

Now, we will examine a composition of inversions that always results in a new inversion. Concretely, we consider two inversions $I_1 = I_{[O,R_1^2]}$ and $I_2 = I_{[Q,R_2^2]}$ in the circles $\mathcal{C}(O, R_1)$ and $\mathcal{C}(Q, R_2)$. Then, the composition

$$I_1 \circ I_2 \circ I_1 = I_{[O,R_1^2]} \circ I_{[Q,R_2^2]} \circ I_{[O,R_1^2]}$$

is a new inversion and we will find its circle and power. We have run into such a composition in this section, and therefore we owe an elaboration about it. The composition, as aforementioned, maps circles to circles or straight lines, and straight lines to circles or straight lines, and since it contains three inversions, it reverses the oriented angles but preserves their absolute values. It is also straightforward to see that *this kind of composition is involutory.* Next, we find a circle that is point-wise invariant under this composition and we prove that the composition is equal to the inversion in this circle.

We study the general case of the above composition with **Figure 4.29**. The other figures that follow after, are special cases depending on the relative position of the two given circles.

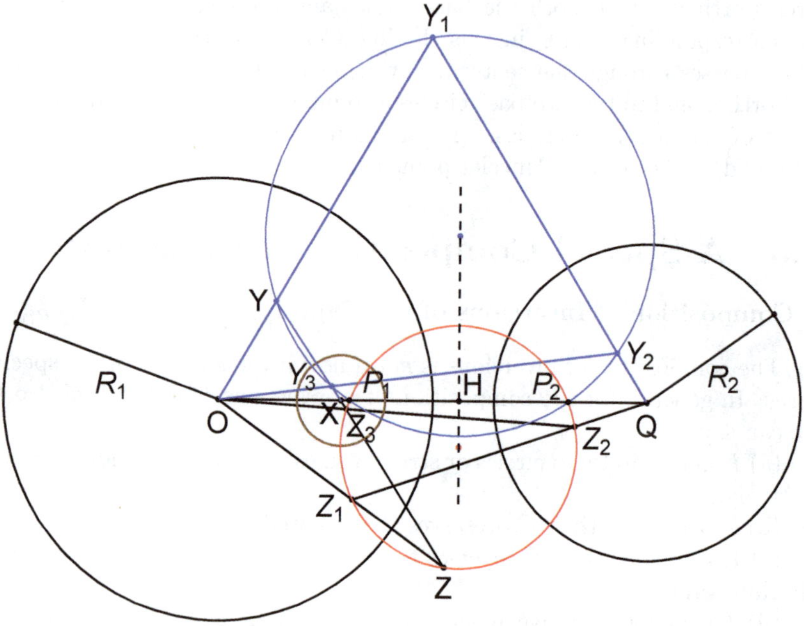

Figure 4.29: Composition of Inversions of the Type $I_{[O,R_1^2]} \circ I_{[Q,R_2^2]} \circ I_{[O,R_1^2]}$

We consider any point Y in the plane and we let

$$Y_1 = I_{[O,R_1^2]}(Y), \qquad Y_2 = I_{[Q,R_2^2]}(Y_1), \qquad Y_3 = I_{[O,R_1^2]}(Y_2).$$

That is, $Y \overset{I_{[O,R_1^2]}}{- - - \longrightarrow} Y_1 \overset{I_{[Q,R_2^2]}}{- - - \longrightarrow} Y_2 \overset{I_{[O,R_1^2]}}{- - - \longrightarrow} Y_3$

$$\text{or} \qquad Y_3 = I_{[O,R_1^2]} \circ I_{[Q,R_2^2]} \circ I_{[O,R_1^2]}(Y).$$

Under the hypotheses

$$\overline{OY} \cdot \overline{OY_1} = R_1^2 = \overline{OY_2} \cdot \overline{OY_3},$$

the four points Y, Y_1, Y_2 and Y_3 are concyclic. Since $Y_1 = I_{[O,R_1^2]}(Y)$ and $Y_3 = I_{[O,R_1^2]}(Y_2)$, by the properties of inversions (see bibliography, or **Section 4.4**), the circle $[YY_1Y_2Y_3]$ is orthogonal to both circles $\mathcal{C}(O, R_1)$ and $\mathcal{C}(Q, R_2)$. So, if we invert the ensuing figure under the inversion $I_{[O,R_1^2]}$, the circle $[YY_1Y_2Y_3]$ is set-wise invariant and the circle $\mathcal{C}(Q, R_2)$ inverts to a circle $\mathcal{C}(X, R)$, of center X and radius R, which is also homothetic to circle $\mathcal{C}(Q, R_2)$, under the homothety with center O and homothetic ratio

$$\lambda = \frac{c_1}{\mathfrak{P}_{\mathcal{C}(Q,R_2)}(O)} = \frac{R_1^2}{OQ^2 - R_2^2}.$$

{Also, Y_1 is mapped to Y and Y_2 to Y_3, under the inversion $I_{[O,R_1^2]}$.}

Now, we consider the circle

$$\mathcal{C}(X, R) := I_{[O,R_1^2]}[\mathcal{C}(Q, R_2)].$$

Then, we notice that:

(a) This circle is fixed under the composition

$$I_{[O,R_1^2]} \circ I_{[Q,R_2^2]} \circ I_{[O,R_1^2]}.$$

(b) The center X is on the center line OQ.

(c)

$$\overline{OX} = \frac{R_1^2 \cdot \overline{OQ}}{OQ^2 - R_2^2}.$$

(d)

$$R = \frac{R_1^2 R_2}{|OQ^2 - R_2^2|}.$$

Now, by the "*invariance of inversiveness under inversion*" (see for instance: Johnson, 2007, 80. Theorem, p.55, Roussos, 2022, Theorem 6.5.4, pp. 354, or **Chapter 4, Theorem 4.4.6.**), the point Y_3 is the inverse point of Y under the inversion $I_{[X,R^2]}$. Therefore,

$$I_{[O,R_1^2]} \circ I_{[Q,R_2^2]} \circ I_{[O,R_1^2]} = I_{[X,R^2]}$$

Remark: All circles of the type (Y, Y_1, Y_2, Y_3), (Z, Z_1, Z_2, Z_3), etc., (as in **Figure 4.29**), are orthogonal to the three circles $\mathcal{C}(O, R_1)$, $\mathcal{C}(Q, R_2)$ and $\mathcal{C}(X, R)$. Therefore: (1) The three circles $\mathcal{C}(O, R_1)$, $\mathcal{C}(Q, R_2)$ and

$\mathcal{C}(X, R)$ have the same radical axis (are coaxial). (2) All circles of the type (Y, Y_1, Y_2, Y_3), (Z, Z_1, Z_2, Z_3), etc., pass through the Poncelet points P_1 and P_2 of the two initially given circles $\mathcal{C}(O, R_1)$ and $\mathcal{C}(Q, R_2)$, if the given circles do not intersect at two points. (3) The centers of all circles (Y, Y_1, Y_2, Y_3), (Z, Z_1, Z_2, Z_3), etc., are on the radical axis of $\mathcal{C}(O, R_1)$ and $\mathcal{C}(Q, R_2)$. [The Poncelet points exist, if the two circles are not concentric, in which case the radical axis is the straight line at infinity. If the circles are tangent, then both Poncelet points coincide at the point of tangency, and if they intersect, the Poncelet points are the two points of intersection. When the circles are concentric, then the circles (Y, Y_1, Y_2, Y_3), etc., become straight lines. See the figures that follow.]

Below, we provide several figures of this composition depending on the relative positions of the circles $\mathcal{C}(O, R_1)$ and $\mathcal{C}(Q, R_2)$. In **Figure 4.30**, the circle $\mathcal{C}(Q, R_2)$ passes through O. Then, we have that the circle

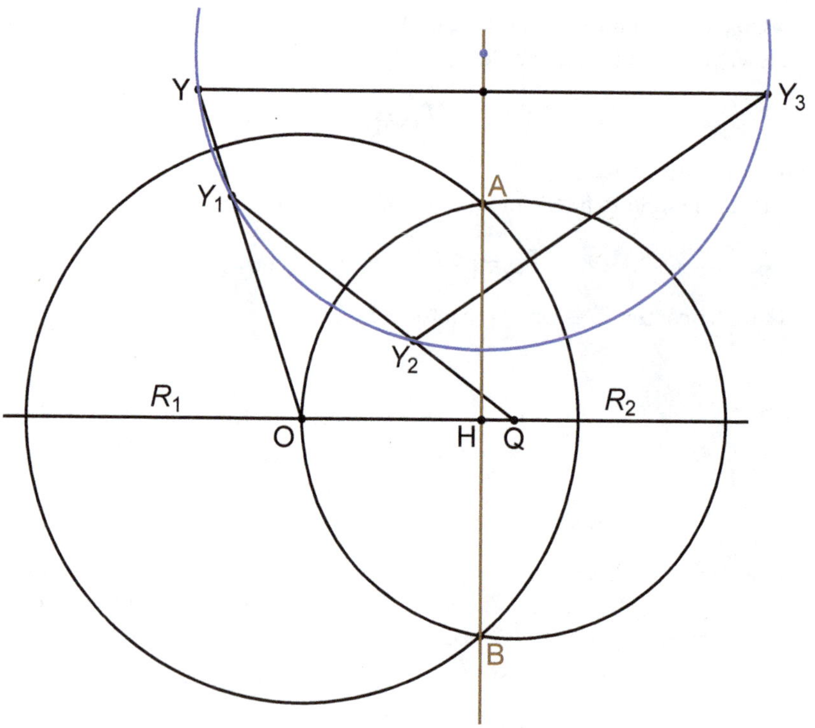

Figure 4.30: Composition of Inversions of Type $I_{[O,R_1^2]} \circ I_{[Q,R_2^2]} \circ I_{[O,R_1^2]}$

$$\mathcal{C}(X,R) := I_{[O,R_1^2]}[\mathcal{C}(Q,R_2)]$$

is a straight line (circle with center at infinity), and the composition reduces to a **reflection** in this line. This straight line is the radical axis of $\mathcal{C}(O,R_1)$ and $\mathcal{C}(Q,R_2)$, which, in this **special case**, is the common chord AB of $\mathcal{C}(O,R_1)$ and $\mathcal{C}(Q,R_2)$ (and the Poncelet points are the points A and B).

The remaining figures explain the **special cases** of this composition relative to the positions of the two circles of the given inversions.

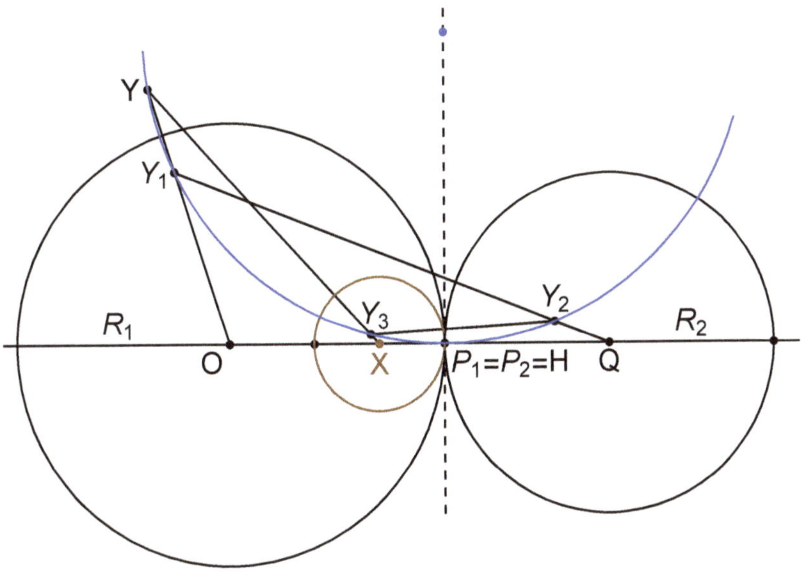

Figure 4.31: Composition of Inversions of Type
$I_{[O,R_1^2]} \circ I_{[Q,R_2^2]} \circ I_{[O,R_1^2]}$

Remarks: (1) Using the above result, we inductively prove that any composition of an odd number of inversions of the form

$$I_n \circ \ldots \circ I_2 \circ I_1 \circ I \circ I_1 \circ I_2 \ldots \circ I_n \quad \text{is an inversion.}$$

(2) If inversions I_1 and I_2 **commute** (iff their circles are orthogonal), then

$$I_2 \circ I_1 \circ I_2 = I_1.$$

(3) For the analytical proof of these results and approach to these figures see the last part of **Section 4.7**.

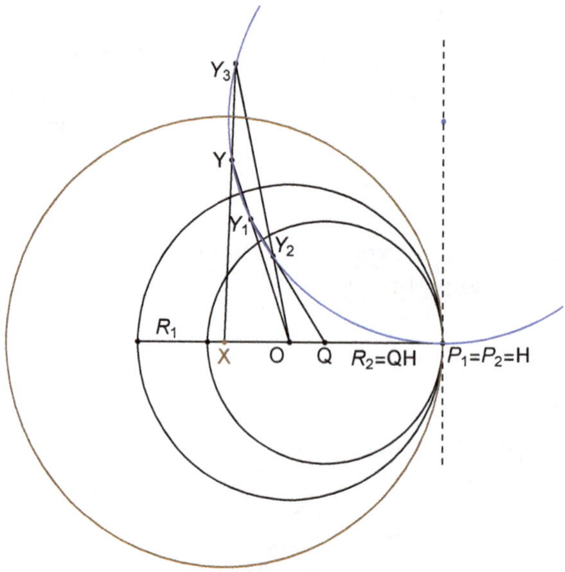

Figure 4.32: Composition of Inversions of Type
$I_{[O,R_1^2]} \circ I_{[Q,R_2^2]} \circ I_{[O,R_1^2]}$

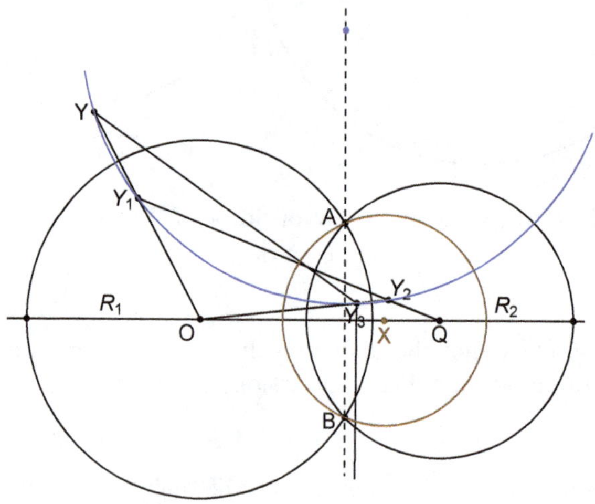

Figure 4.33: Composition of Inversions of Type
$I_{[O,R_1^2]} \circ I_{[Q,R_2^2]} \circ I_{[O,R_1^2]}$

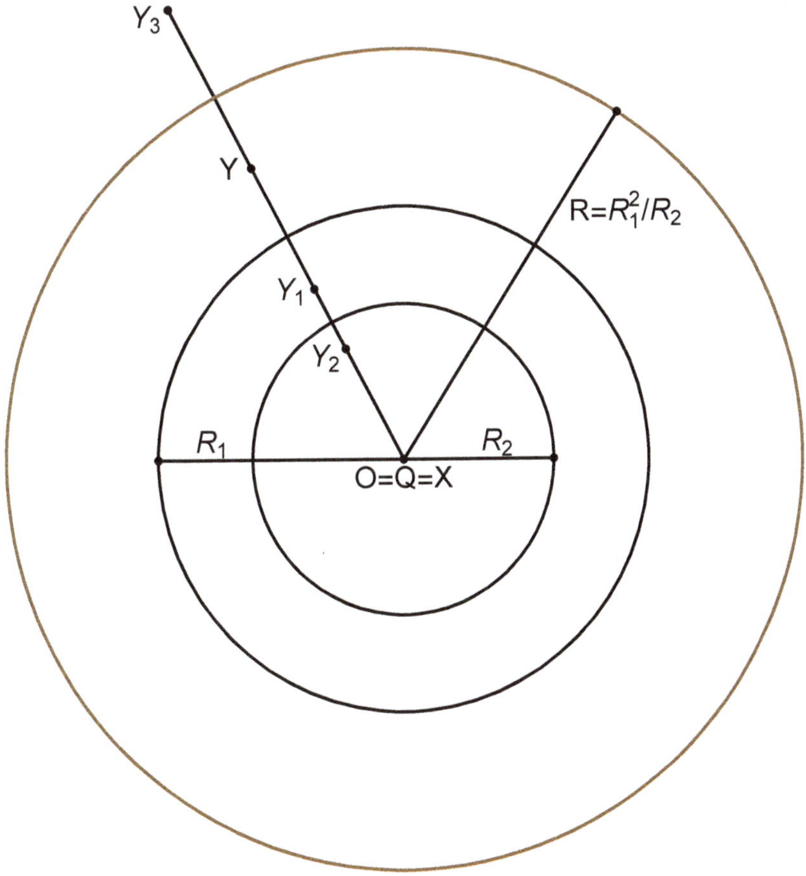

Figure 4.34: Composition of Inversions of Type $I_{[O,R_1^2]} \circ I_{[Q,R_2^2]} \circ I_{[O,R_1^2]}$

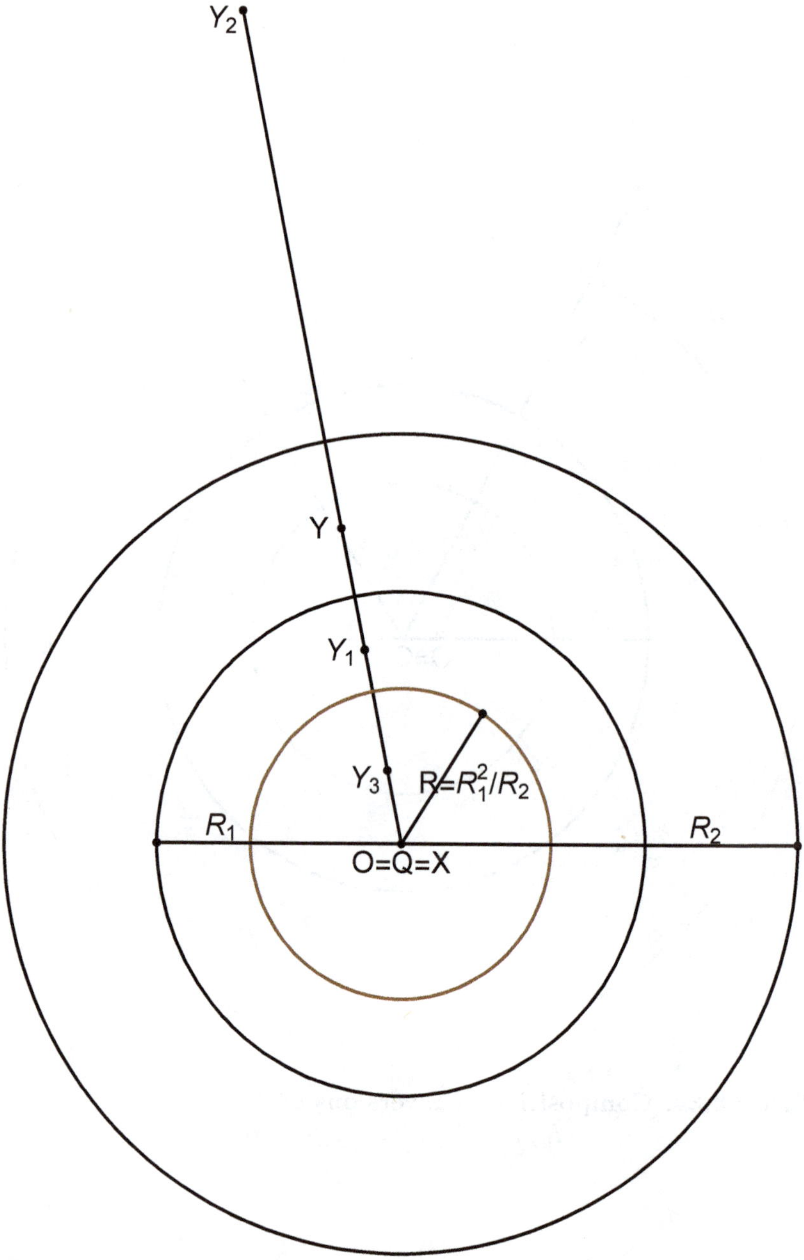

Figure 4.35: Composition of Inversions of Type
$I_{[O,R_1^2]} \circ I_{[Q,R_2^2]} \circ I_{[O,R_1^2]}$

4.6 Descartes's Theorem for Tangent Circles.

In **Figure 4.36**, we are given three circles (A, r_1), (B, r_2), and (C, r_3) none of which is inside another and whose centers are not on a straight line. Considering the two types of tangency of two circles (internal or external), it is known that there are $8 = 2^3$ circles that can be tangent to these three circles, in general (but not always).[2] (See, for example, Pedoe D., *Geometry a Comprehensive Course*, pp. 101-104, and other bibliography.)

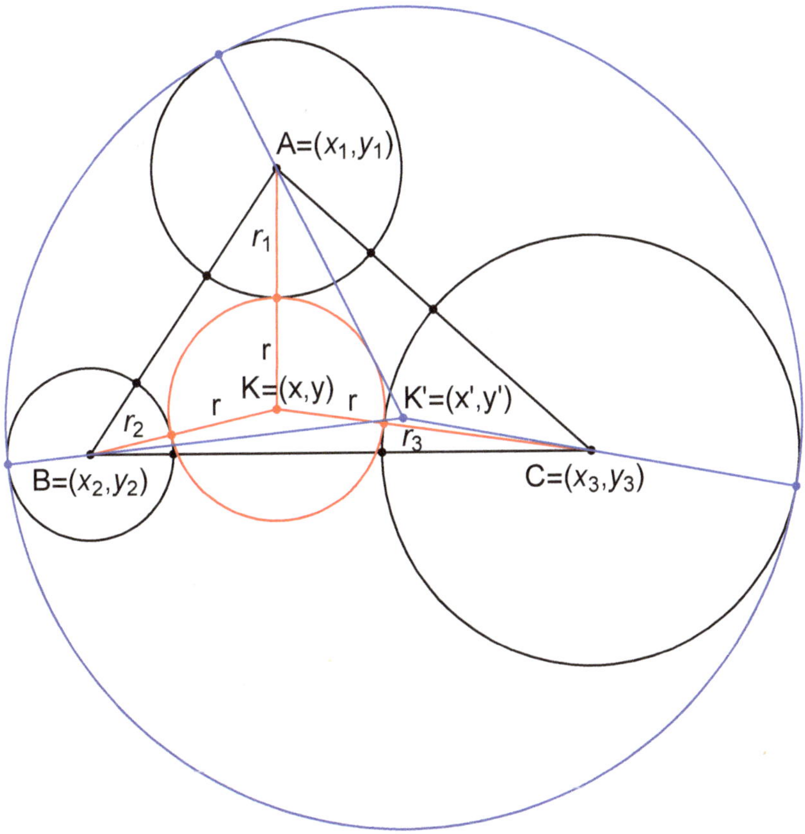

Figure 4.36: Three Circles Pairwise Outside Each Other with Two Circles Tangent to Them.

[2]This result is a partial solution to the so called **Apollonius Problem of Tangencies**, which has many cases and subcases. (Apollonius of Perga, Greek mathematician, c.262-c.190 BCE.)

In **Figure 4.36**, we have depicted two of the circles, the smaller one (K, r) (red) externally tangent to the three circles and the larger one (K', r') (blue) internally tangent to them. So, if we let $d_1 = BC$, $d_2 = CA$ and $d_3 = AB$, in this case, we have

$$d_1 > r_2 + r_3, \quad \text{and} \quad d_2 > r_3 + r_1, \quad \text{and} \quad d_3 > r_1 + r_2.$$

In this case, the radius of a given circle could be zero, that is the circle is a point. If all three circles are points, then we look for the circle that passes through them.

The question is how we can find the radii r and r'. Finding these radii, the center K is the common point of the three circles $(A, r_1 + r)$, $(B, r_2 + r)$, and $(C, r_3 + r)$, and K' is the common point of the three circles $(A, r' - r_1)$, $(B, r' - r_2)$, and $(C, r' - r_3)$. (Analogous adjustments for the other 6 circles.)

Deriving a formula (analogous to the formula in Descartes' Theorem) for the radii r and r', in this general case, is not easy. One way is to use the Law of Cosines for the angles formed around the centers K and K', which add up to 2π. For instance, if we consider the angle $a := \widehat{BKC}$ we have

$$\cos(a) = \frac{(r + r_2)^2 + (r + r_3)^2 - d_1^2}{2(r + r_2)(r + r_3)} \tag{4.5}$$

and so on, for the angles $b := \widehat{CKA}$ and $c := \widehat{AKB}$.

We observe that $a + b + c = 2\pi$, and for such angles, the identity

$$\cos^2(a) + \cos^2(b) + \cos^2(c) = 1 + 2\cos(a)\cos(b)\cos(c) \tag{4.6}$$

is valid (prove it). Plugging in the above expressions of cosines and solving for r, we derive a formula for r. In this case the formula obtained is very lengthy and inconvenient for use. Plugging in the numerical values of r_1, r_2, r_3, d_1, d_2, and d_3, we can solve the equation numerically and find r.

A convenient way to find K and r is by solving the following system, analytically or numerically. We use coordinates and let $K = (x, y)$, $A = (x_1, y_1)$, $B = (x_2, y_2)$, and $C = (x_3, y_3)$. Then, we must solve the system:

$$\begin{aligned}(x - x_1)^2 + (y - y_1)^2 &= (r + r_1)^2\\(x - x_2)^2 + (y - y_2)^2 &= (r + r_2)^2\\(x - x_3)^2 + (y - y_3)^2 &= (r + r_3)^2\end{aligned}$$

To find $K = (x', y')$ and r', we replace x, y, r by x', y', r' and the $+r_1, +r_2, +r_3$ by $-r_1, -r_2, -r_3$ and solve the new system. A good computer package can solve this system efficiently and accurately.

However, in **Descartes' case**, the three circles (A, r_1), (B, r_2), and (C, r_3), must have radii > 0, (i.e., no circle is a point) and are mutually externally tangent. Thus, $d_1 = r_2 + r_3$, $d_2 = r_3 + r_1$, and $d_3 = r_1 + r_2$. Using these relations in **equations 4.5** and **4.6** and simplifying, we derive the formula for the circle (K, r)

$$2\left(\frac{1}{r_1^2} + \frac{1}{r_2^2} + \frac{1}{r_3^2} + \frac{1}{r^2}\right) = \left(\frac{1}{r_1} + \frac{1}{r_2} + \frac{1}{r_3} + \frac{1}{r}\right)^2 \iff$$

$$\frac{1}{r_1^2} + \frac{1}{r_2^2} + \frac{1}{r_3^2} + \frac{1}{r^2} = 2\left(\frac{1}{r_1}\frac{1}{r_2} + \frac{1}{r_1}\frac{1}{r_3} + \frac{1}{r_1}\frac{1}{r} + \frac{1}{r_2}\frac{1}{r_3} + \frac{1}{r_2}\frac{1}{r} + \frac{1}{r_3}\frac{1}{r}\right)$$

suggested by Descartes. (Try it out! See **Case (1), item (6)** and **Case (2), item (5)** of **Section 2.4**.) That is, this method proves **Descartes's Theorem** using some trigonometry. Solving this equation for r, we find two solutions that we must investigate.

The Theorem is also valid, if one or two of the circles are straight line(s), in which case, we consider two straight lines to be tangent iff they are parallel, and in the formula we put $\frac{1}{\infty} = 0$, for each line. Also, we need adjustments of the signs. If for the external tangency we use the $+$ sign, then for the internal we must use the $-$ sign. (See also **Chapter 4, Theorem 4.6.1**.)

Another way to prove **Descartes's formula**, is to use Heron's formula for the area of a triangle. We have that the point K is interior to the triangle ABC, in this case, and so,

$$\text{area } (ABC) = \text{area } (KBC) + \text{area } (KCA) + \text{area } (KAB).$$

Then, applying Heron's formula, we find

$$\sqrt{r_1 r_2 r_3 (r_1 + r_2 + r_3)} =$$
$$= \sqrt{r_2 r_3 r (r_2 + r_3 + r)} + \sqrt{r_3 r_1 r (r_3 + r_1 + r)} + \sqrt{r_1 r_2 r (r_1 + r_2 + r)}.$$

Manipulating this equation carefully, through squaring and simplifying in every step, we can derive Descartes's formula, as stated above.

We can also extend this method for the other variations of this formula, where some circles are externally and other internally tangent. See also the pertinent bibliography.

Now, we are going to prove the Theorem of Descartes using inversion.

Theorem 4.6.1 (René Descartes) *Each circle out of four circles (O_i, r_i), with $r_i > 0$, $i = 1, 2, 3, 4$, in the plane is externally tangent to the other three circles if and only if the following relation among the radii holds:*

$$2\left(\frac{1}{r_1^2} + \frac{1}{r_2^2} + \frac{1}{r_3^2} + \frac{1}{r_4^2}\right) = \left(\frac{1}{r_1} + \frac{1}{r_2} + \frac{1}{r_3} + \frac{1}{r_4}\right)^2, \qquad (4.7)$$

which is equivalent to

$$\frac{1}{r_1^2}+\frac{1}{r_2^2}+\frac{1}{r_3^2}+\frac{1}{r_4^2}=2\left(\frac{1}{r_1r_2}+\frac{1}{r_1r_3}+\frac{1}{r_1r_4}+\frac{1}{r_2r_3}+\frac{1}{r_2r_4}+\frac{1}{r_3r_4}\right). \quad (4.8)$$

The quantities $\kappa_i := \frac{1}{r_i}$, $i = 1, 2, 3, 4$, are called **curvatures** (or **bends**) of the circles and using this notation, we can write the above relation[3] as

$$2(\kappa_1^2+\kappa_2^2+\kappa_3^2+\kappa_4^2)=(\kappa_1+\kappa_2+\kappa_3+\kappa_4)^2,$$

which is equivalent to

$$\kappa_1^2+\kappa_2^2+\kappa_3^2+\kappa_4^2=2(\kappa_1\kappa_2+\kappa_1\kappa_3+\kappa_1\kappa_4+\kappa_2\kappa_3+\kappa_2\kappa_4+\kappa_3\kappa_4).$$

(The Theorem is a special case of the Apollonius problem of tangencies among four circle, some of which can be points or straight line, in general.).

Proof Let us consider four circles (O_i, r_i), $i = 1, 2, 3, 4$, in the plane that are externally tangent. Let circles (O_1) and (O_2) be tangent at the point T, and assume, without loss of generality, $r_1 \leq r_2$. See **Figure 4.37**.

We are going to apply an inversion with **center** T and convenient **power** $c = 4r_1r_2$. Then, the circles (O_1) and (O_2) are transformed to the parallel lines l_1 and l_2 perpendicular to the center line O_1O_2 at the points P' and Q', inverses of the points P and Q, as in the figure. The circles (O_3) and (O_4) are transformed to two externally tangent circles (O_3') and (O_4') and tangent to the parallel lines l_1 and l_2, and therefore they are equal.

We have: $TP = 2r_1$ and $TP \cdot TP' = c = 4r_1r_2$. Therefore, $TP' = 2r_2$. Similarly, $TQ' = 2r_1$. So, $P'Q' = 2(r_1 + r_2)$. Then, the radii of the circles (O_3') and (O_4') are equal to

$$r_3' = r_4' = \frac{P'Q'}{2} = r_1 + r_2.$$

We draw the parallel radii $O_3'A = r_1 + r_2$ and $O_4' = r_1 + r_2$ and the segment $TA'B'$ perpendicular to them,as in the figure. We then have $AA' = BB' = 2r_1$, $O_3'A' = O_4'B' = r_1 + r_2 - 2r_1 = r_2 - r_1$, and $A'B' = AB = 2(r_1 + r_2)$.

We also need to compute $x := Q'A$. We have that

$$TO_3'^2 = x^2 + O_3'A'^2 = x^2 + (r_2 - r_1)^2, \quad \text{and}$$
$$TO_4'^2 = [Q'A' + A'B']^2 + O_4'B'^2 = [x + 2(r_1 + r_2)]^2 + (r_2 - r_1)^2.$$

[3]This relation was given by Frederick Soddy, (F R S, 2 September 1877 – 22 September 1956, English radiochemist). R. Descartes knew the Theorem but did not write this relation in this way. Also, others rediscovered the Theorem after Descartes. There is a lot of interesting history behind this Theorem.

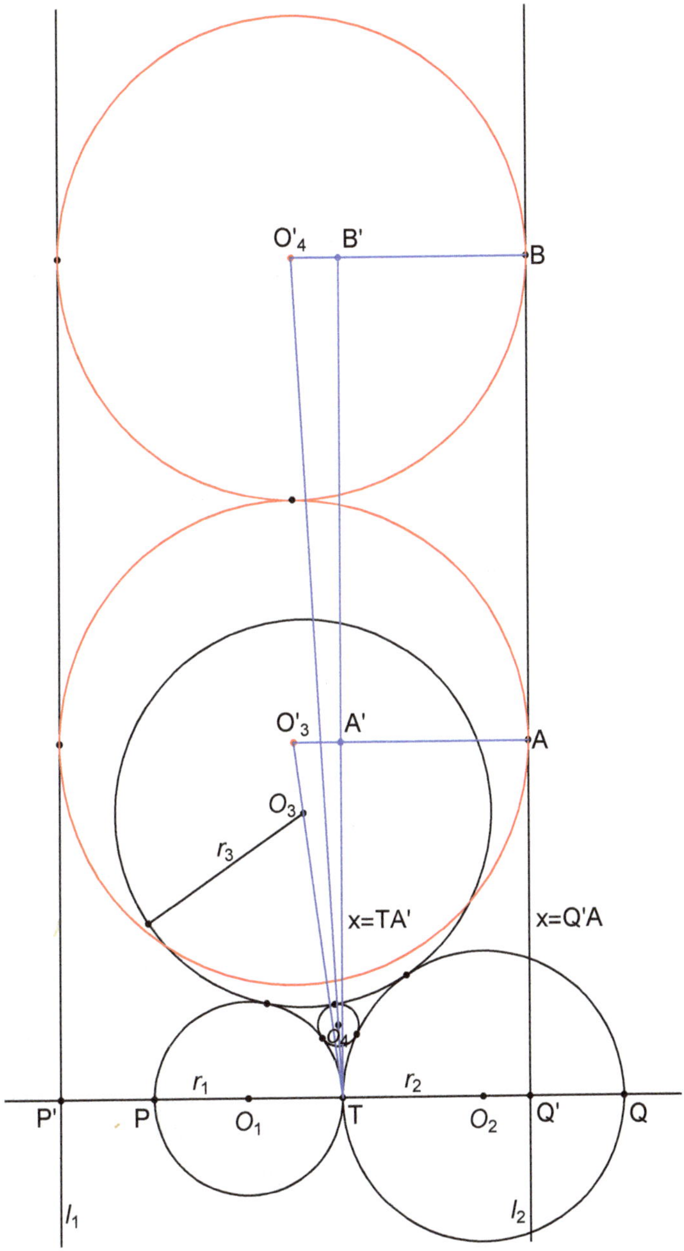

Figure 4.37: For the Proof of the Theorem of Descartes by Inversion

The powers of T with respect to the circles (O_3', r_1+r_2) and (O_4', r_1+r_2) are

$$\mathfrak{P}_3' = TO_3'^2 - (r_1+r_2)^2 = x^2 + (r_2-r_1)^2 - (r_1+r_2)^2 = x^2 - 4r_1r_2, \quad \text{and}$$

$$\mathfrak{P}_4' = TO_4'^2 - (r_1+r_2)^2 =$$
$$[x+2(r_1+r_2)]^2 + (r_2-r_1)^2 - (r_1+r_2)^2 = [x+2(r_1+r_2)]^2 - 4r_1r_2.$$

The circles (O_3) and (O_3') are inverse with respect to used the inversion and so they are homothetic. We consider O_3 to be the homothetic image of (O_3'), under the homothety with center T and homothetic ratio

$$\lambda_3 = \frac{c}{\mathfrak{P}_3'} = \frac{4r_1r_2}{x^2 - 4r_1r_2}.$$

Then, comparing the radii, we get

$$\frac{r_3}{r_1+r_2} = \lambda_3 = \frac{c}{\mathfrak{P}_3'} = \frac{4r_1r_2}{x^2 - 4r_1r_2}.$$

Solving this for x^2, we find

$$x^2 = \frac{4r_1r_2(r_1+r_2+r_3)}{r_3} \Longrightarrow x = \sqrt{\frac{4r_1r_2(r_1+r_2+r_3)}{r_3}}.$$

Then,

$$\mathfrak{P}_4' = [x+2(r_1+r_2)]^2 - 4r_1r_2 =$$
$$\left[\sqrt{\frac{4r_1r_2(r_1+r_2+r_3)}{r_3}} + 2(r_1+r_2)\right]^2 - 4r_1r_2 =$$
$$= \frac{4r_1r_2(r_1+r_2+r_3)}{r_3} + 8(r_1+r_2)\sqrt{\frac{r_1r_2(r_1+r_2+r_3)}{r_3}} +$$
$$+4(r_1+r_2)^2 - 4r_1r_2.$$

Also, the circles (O_4) and (O_4') are inverse with respect to used the inversion and so they are homothetic. We consider O_4 to be the homothetic image of (O_4'), under the homothety with center T and homothetic ratio

$$\lambda_4 = \frac{c}{\mathfrak{P}_4'} = \frac{4r_1r_2}{\mathfrak{P}_4'}.$$

Then, comparing the radii, we get

$$\frac{r_4}{r_1+r_2} = \lambda_4 = \frac{4r_1r_2}{\mathfrak{P}_4'} \Longrightarrow \frac{1}{r_4} = \frac{\mathfrak{P}_4'}{4r_1r_2(r_1+r_2)}.$$

We substitute the $\mathfrak{P}'_4$ (as found above) and we manipulate the obtained relation until we find

$$\frac{1}{r_4} = \frac{1}{r_1} + \frac{1}{r_2} + \frac{1}{r_3} + 2\sqrt{\frac{1}{r_2 r_3} + \frac{1}{r_1 r_3} + \frac{1}{r_1 r_2}},$$

or

$$\frac{1}{r_4} - \frac{1}{r_1} - \frac{1}{r_2} - \frac{1}{r_3} = 2\sqrt{\frac{1}{r_2 r_3} + \frac{1}{r_1 r_3} + \frac{1}{r_1 r_2}}.$$

Squaring this equality and simplifying, we get

$$\frac{1}{r_1^2} + \frac{1}{r_2^2} + \frac{1}{r_3^2} + \frac{1}{r_4^2} = 2\left(\frac{1}{r_1 r_2} + \frac{1}{r_1 r_3} + \frac{1}{r_1 r_4} + \frac{1}{r_2 r_3} + \frac{1}{r_2 r_4} + \frac{1}{r_3 r_4}\right),$$

which is the relation that we wanted to prove.

■

In the configuration of **Figure 4.37**, the relation is essentially

$$\frac{1}{r_4} = \frac{1}{r_1} + \frac{1}{r_2} + \frac{1}{r_3} + 2\sqrt{\frac{1}{r_2 r_3} + \frac{1}{r_1 r_3} + \frac{1}{r_1 r_2}},$$

that computes the radius r_4 in terms of the other three radii. But, if we solve the squared final relation

$$2\left(\frac{1}{r_1^2} + \frac{1}{r_2^2} + \frac{1}{r_3^2} + \frac{1}{r_4^2}\right) = \left(\frac{1}{r_1} + \frac{1}{r_2} + \frac{1}{r_3} + \frac{1}{r_4}\right)^2$$

for r_4, we also find an additional solution, namely

$$\frac{1}{r_4} = \frac{1}{r_1} + \frac{1}{r_2} + \frac{1}{r_3} - 2\sqrt{\frac{1}{r_2 r_3} + \frac{1}{r_1 r_3} + \frac{1}{r_1 r_2}}.$$

This solution has also geometrical meaning. Its absolute value is the radius of the circle that it tangent internally to the circles (O_1), (O_2), and (O_3), whereas any one of them is externally tangent to the other two circles. **Figure 4.38.**

This absolute value is also obtained if we replace $\frac{1}{r_4}$ by $-\frac{1}{r_4}$ in **equation 4.7**, or **4.8**, and take the absolute value of the solution with the $+$ in front of the square root. In general, we adjust Descartes' Theorem by considering the external tangencies with $+$ and the internal with $-$ and in the end we take the appropriate solution with absolute value, if necessary.

Working in analogous way, as in the proof above, with **Figure 4.38**, we derive the second solution for r_4. In fact we have:

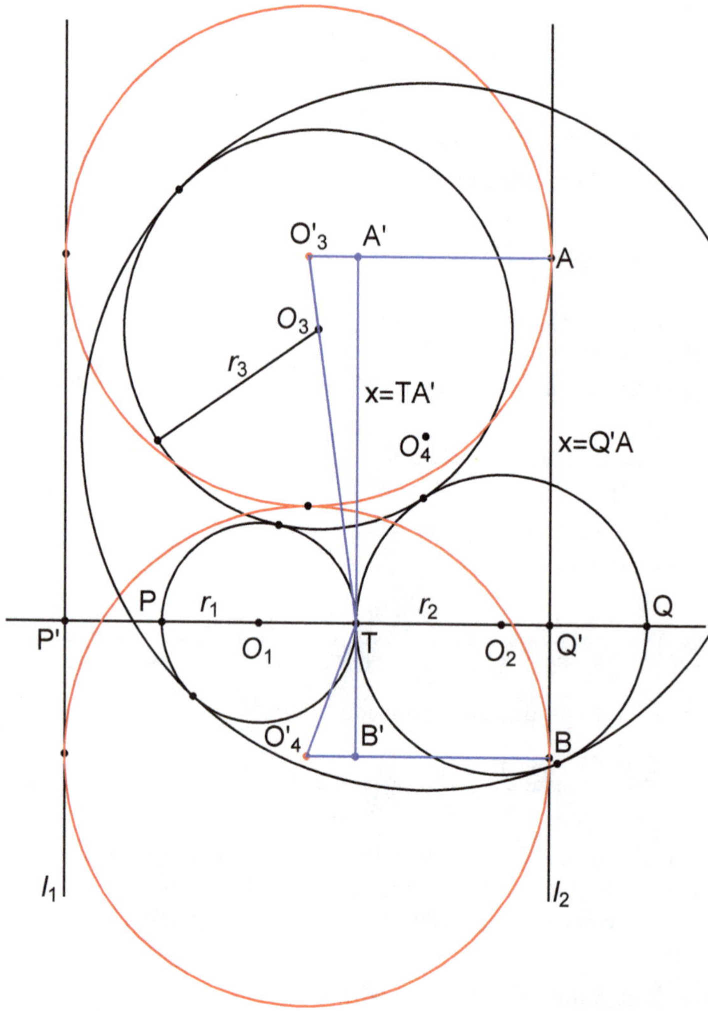

Figure 4.38: For the Proof of the Theorem of Descartes by Inversion

(a) The results concerning the circles (O_3) and (O_3') remain the same.
(b) Now we have the power

$$\mathfrak{P}_4' = \frac{4r_1r_2(r_1+r_2+r_3)}{r_3} - 8(r_1+r_2)\sqrt{\frac{r_1r_2(r_1+r_2+r_3)}{r_3}} + \\ +4(r_1+r_2)^2 - 4r_1r_2.$$

Using this, we get

$$\frac{1}{r_4} = \frac{1}{r_1} + \frac{1}{r_2} + \frac{1}{r_3} - 2\sqrt{\frac{1}{r_2r_3} + \frac{1}{r_1r_3} + \frac{1}{r_1r_2}}.$$

Note: The Theorem of Descartes is also valid if one or two of the circles are straight lines. For a straight line $r = \infty$, and so, $\frac{1}{r} = 0$. The **curvature (or bend) of a straight line is zero**. We keep in mind that two straight lines are considered to be tangent iff they are parallel. (Elaborate these cases. Things become simpler.)

4.7 Analytical Approach to Inversion

Suppose we work in the cartesian plane with an orthonormal system of axes and the points are represented by pairs of coordinates (x, y). We consider an inversion $\mathfrak{I}$, with center the point $O = (h, k)$ and power $c \neq 0$. Given any point $P = (x, y) \neq O = (h, k)$, we want to find its inverse $P' = (x', y')$. [The inverse of the point $O = (h, k)$ vanishes at infinity.]

Since $\overline{OP} \cdot \overline{OP'} = c$, we have $\overline{OP'} = \frac{c}{\overline{OP}}$. Then,

$$\|\overline{OP'}\| = \frac{|c|}{\|\overline{OP}\|},$$

or

$$\sqrt{(x'-h)^2 + (y'-k)^2} = \frac{|c|}{\sqrt{(x-h)^2 + (y-k)^2}}.$$

Also, $\overline{OP}$ and $\overline{OP'}$ have the same or opposite direction depending on $c > 0$ or $c < 0$, respectively. Since any nonzero vector is equal to the unit vector of the same direction multiplied by the length of the given vector, we get

$$(x'-h, y'-k) = \frac{\pm|c|}{\sqrt{(x-h)^2 + (y-k)^2}} \cdot \frac{(x-h, y-k)}{\sqrt{(x-h)^2 + (y-k)^2}},$$

or

$$(x'-h, y'-k) = \frac{c \cdot (x-h, y-k)}{(x-h)^2 + (y-k)^2}.$$

Therefore,

$$(x', y') = (h, k) + \frac{c(x-h, y-k)}{(x-h)^2 + (y-k)^2} \quad \Longleftrightarrow$$

$$\begin{cases} x' = h + \dfrac{c(x-h)}{(x-h)^2 + (y-k)^2} \\ \\ y' = k + \dfrac{c(y-k)}{(x-h)^2 + (y-k)^2}. \end{cases} \tag{4.9}$$

[Also, solving this system for x and y, we find

$$\begin{cases} x = h + \dfrac{c(x'-h)}{(x'-h)^2 + (y'-k)^2} \\ \\ y = k + \dfrac{c(y'-k)}{(x'-h)^2 + (y'-k)^2}, \end{cases}$$

as it should be, because P' is the inverse of P.]

If we work in the complex plane, then the above formulae are written more elegantly with complex numbers. Let $O = w$, $P = z$ and $P' = z'$. Then we have,

$$z' = w + \frac{c(z-w)}{|z-w|^2} = w + \frac{c(z-w)}{(z-w)(\overline{z-w})} =$$

$$= w + \frac{c}{\overline{z-w}} = w + \frac{c}{\overline{z} - \overline{w}}.$$

(If we solve this for z, we find the inverse relation, which is the similar equation

$$z = w + \frac{c}{\overline{z'-w}} = w + \frac{c}{\overline{z'} - \overline{w}}.$$

Inversions are anti-conformal mappings in the complex plane, that is, their conjugates are conformal.)

Inverses of Straight Lines.

The general cartesian equation of a straight line l in the real cartesian $x-y$ plane is

$$ax + by + c = 0,$$

where a and b are real scalar constants, such that, $|a| + |b| \neq 0$, and c is any real constant.

To find the image of this straight line in the same plane, under the inversion $\mathfrak{I}$, with center the point $O = (h, k)$ and power $c = r^2 > 0$, we need to replace the x and y of this equation, by the x' and y' found in **equations 4.9**. So, we get

$$a\left[h + \frac{r^2(x-h)}{(x-h)^2 + (y-k)^2}\right] + b\left[k + \frac{r^2(y-k)}{(x-h)^2 + (y-k)^2}\right] + c = 0. \quad (4.10)$$

This equation, is equivalent to

$$(ah + bk + c)\left[(x-h)^2 + (y-k)^2\right] + r^2ax + r^2by - r^2(ah + bk) = 0. \quad (4.11)$$

So, there are two cases:

<u>**Case 1:**</u> $ah + bk + c = 0.$

That is, in this case, the center of the inversion (h, k) belongs to the straight line l. Then, canceling r^2 from **equation 4.11**, we obtain

$$ax + by - (-c) = ax + by + c = 0,$$

which is the same straight line as the initial one l, as expected.

<u>**Case 2:**</u> $ah + bk + c \neq 0.$

Then, **equation 4.11** is a circle, whose center and radius we must find. Dividing by $ah + bk + c \neq 0$ and completing the squares, we find

$$\left[x - \left(h - \frac{r^2a}{2(ah + bk + c)}\right)\right]^2 + \left[y - \left(k - \frac{r^2b}{2(ah + bk + c)}\right)\right]^2 =$$
$$= \frac{r^4\left(a^2 + b^2\right)}{4(ah + bk + c)^2}. \quad (4.12)$$

This is a circle with center

$$K = \left[h - \frac{r^2a}{2(ah + bk + c)}, k - \frac{r^2b}{2(ah + bk + c)}\right] =$$
$$= (h, k) - \frac{r^2}{2(ah + bk + c)}(a, b)$$

and radius

$$R = \frac{r^2\sqrt{a^2+b^2}}{2|ah+bk+c|}.$$

Notice that

$$d := \frac{|ah+bk+c|}{\sqrt{a^2+b^2}}$$

is the distance of the center of the inversion (h, k) from the line l, and so,

$$R = \frac{r^2}{2d}.$$

Also, for $x = h$ and $y = k$ **equation 4.12** is satisfied. That is, **circle 4.12** (the inverse image of l) passes through the center of inversion, as expected.

The straight line through $O = (h, k)$ and perpendicular to l has equation $b(x-h) - a(y-k) = 0$. The center K, found above, satisfies this equation. Therefore, this straight line passes through the center of **circle 4.12** and intersects it at a point diametrical to $O = (h, k)$. This point is

$$Q = \left(h - \frac{r^2 a}{ah+bk+c}, k - \frac{r^2 b}{ah+bk+c}\right).$$

Inverses of Circles.

A circle $\mathfrak{C} = (C, R)$ with center $C = (a, b)$ and radius $R > 0$ has general analytic equation

$$(x-a)^2 + (y-b)^2 = R^2,$$

or

$$x^2 + y^2 - 2ax - 2by + a^2 + b^2 - R^2 = 0.$$

To find the image of this circle in the same plane, under the inversion $\mathfrak{I}$, with center the point $O = (h, k)$ and power $c = r^2 > 0$, we need to replace the x and y of this equation, by the x' and y' found in **equation 4.9**. So, we must expand

$$\left[h + \frac{r^2(x-h)}{(x-h)^2+(y-k)^2}\right]^2 + \left[k + \frac{r^2(y-k)}{(x-h)^2+(y-k)^2}\right]^2 -$$
$$-2a\left[h + \frac{r^2(x-h)}{(x-h)^2+(y-k)^2}\right] - 2b\left[k + \frac{r^2(y-k)}{(x-h)^2+(y-k)^2}\right] +$$
$$+a^2 + b^2 - R^2 = 0.$$

This can be transformed to

$$\left[(a-h)^2+(b-k)^2-R^2\right]\left[(x-h)^2+(y-k)^2\right]+ \\ +r^2[2(h-a)(x-h)+2(k-b)(y-k)+r^2]=0. \qquad (4.13)$$

Here, we notice that $\mathfrak{P}=(a-h)^2+(b-k)^2-R^2$ is the power of $O=(h,k)$ with respect to $\mathfrak{C}=(C,R)$.

Again we have two cases:

<u>**Case 1:**</u> $\mathfrak{P}=(a-h)^2+(b-k)^2-R^2=[\text{distance}(O,C)]^2-R^2=0.$

This happens iff the circle $\mathfrak{C}=(C,R)$ passes through the center of the inversion $O=(h,k)$, since $(a-h)^2+(b-k)^2=R^2$. Hence, in this case, the image of $\mathfrak{C}=(C,R)$ under the inversion is the **straight line** given by

$$2(h-a)(x-h)+2(k-b)(y-k)+r^2=0,$$

as expected. [Using $(a-h)^2+(b-k)^2-R^2=0$, the equation of the straight line can also be written $2(h-a)x+2(k-b)x+a^2+b^2+r^2-h^2-k^2-R^2=0$, if this is useful in some cases.]

The distance of the line from $C=(a,b)$ the center of $\mathfrak{C}=(C,R)$ is

$$d=\frac{|-2(h-a)^2-2(k-b)^2+r^2|}{2\sqrt{(h-a)^2+(k-b)^2}}=\frac{|-2R^2+r^2|}{2R}.$$

<u>**Case 2:**</u> $\mathfrak{P}=(a-h)^2+(b-k)^2-R^2=[\text{distance}(O,C)]^2-R^2\neq 0.$

Now, the circle $\mathfrak{C}=(C,R)$ does not pass through the center of inversion, $O=(h,k)$. **Equation 4.13** is the equation of a circle, as expected, whose center and radius must be determined.

Then, we transform (carefully) this equation into the standard equation of the **circle** and we find

$$\left\{x-\left[h+\frac{r^2}{\mathfrak{P}}(a-h)\right]\right\}^2+\left\{y-\left[k+\frac{r^2}{\mathfrak{P}}(b-k)\right]\right\}^2=\frac{r^4}{\mathfrak{P}^2}R^2, \qquad (4.14)$$

where

$$\mathfrak{P}=\mathfrak{P}_{\mathfrak{C}}(O)=(a-h)^2+(b-k)^2-R^2\neq 0.$$

So, the **center** of the image circle is

$$K=\left[h+\frac{r^2}{\mathfrak{P}}(a-h),k+\frac{r^2}{\mathfrak{P}}(b-k)\right],$$

and the **radius** is

$$\rho = \frac{r^2}{|\mathfrak{P}|} R.$$

Equation 4.14 also says that the circle which is the inverse image of the given circle $\mathfrak{C} = (C, R)$ is **homothetic** to the given circle $\mathfrak{C} = (C, R)$, with center of homothety the center of the inversion $\mathfrak{I}$, (h, k), and homothetic ratio $\frac{r^2}{\mathfrak{P}}$ [the power of the inversion r^2 divided by the power of (h, k), the center of the inversion, with respect to the circle $\mathfrak{C} = (C, R)$], as expected.

Let C' be the inverse point of C. Notice that the points O, C, C', and K are collinear. But C' and K are different. The points C and C' are inverse and the points C and K are homothetic. The fact that $C' \neq K$ follows by computing their distances from O and see that they are different. In fact, we have:

$$\overline{OC} \cdot \overline{OC'} = r^2.$$

Therefore,

$$\text{distance}(O, C') = \frac{r^2}{\text{distance}(O, C)} = \frac{r^2}{\sqrt{\mathfrak{P} + R^2}}.$$

Next,

$$\text{distance}(O, K) = \frac{r^2}{|\mathfrak{P}|} \sqrt{\mathfrak{P} + R^2}.$$

Obviously, these two distances are different, except when

$$\mathfrak{P} = [\text{distance}(O, C)]^2 - R^2 = \frac{-R^2}{2} < 0.$$

But then,

$$\text{distance}(O, C) = \frac{R}{\sqrt{2}},$$

and so, the point O is inside the circle $\mathfrak{C} = (C, R)$ at distance $\frac{R}{\sqrt{2}}$ from its center C. In such a case, the homothetic ratio of the related homothety is $\frac{-2r^2}{R^2} < 0$, and so, the homothety is negative. Because of the negativity of the homothety whereas the inversion is positive, we again have $C' \neq K$.

Remark: Suppose the circle of inversion (O, r) and the given circle (C, R) are concentric $(O = C)$. Then, $a = h$, $b = k$, $\mathfrak{P} = -R^2$, and **equation 4.14** takes the form

$$(x - h)^2 + (y - k)^2 = \frac{r^4}{R^2}.$$

Therefore, in this situation, the inverse image of (C, R) is also concentric with (O, r) and (C, R) and its radius is $\frac{r^2}{R}$. If also $R = r$, then the radius of the image circle is r. That is:
The circle (O, r) of a positive inversion (center O and power r^2) is pointwise invariant under the inversion.

Poncelet Points and Radical Axis with Analytic Geometry

Consider two circles $C_1[O_1, R_1] : (x - x_1)^2 + (y - y_1)^2 = R_1^2$ and $C_2[O_2, R_2] : (x - x_2)^2 + (y - y_2)^2 = R_2^2$. Then, the center line O_1O_2 has equation

$$(x_2 - x_1)(y - y_1) = (y_2 - y_1)(x - x_1)$$

Denote the length of the center segment O_1O_2 with

$$k := |O_2 - O_1| = \sqrt{(x_2 - x_1)^2 + (y_2 - y_1)^2} > 0.$$

(If $k = 0$, the circles are concentric and **Equation 3.8** is meaningless.)

We have seen **Equation 3.8** in **Section 3.1** and its derivation in **Section 4.4**. Its solutions of are

$$s_\pm = \frac{k^2 + (R_1 + R_2)(R_1 - R_2)}{2k} \pm$$
$$\pm \frac{\sqrt{(k + R_1 + R_2)(k + R_1 - R_2)(k - R_1 + R_2)(k - R_1 - R_2)}}{2k}.$$

This equation and its solutions thereof were derived assuming that $\overline{O_1O_2}$ was an axis with point of reference O_1.

When the **circles do not intersect**, we have the following three cases.

(1) If $k > R_1 + R_2$ the circles are exterior of one another (and do not intersect). Then the two **Poncelet points**, in this situation, are inside the segment O_1O_2 and are given analytically by

$$P_\pm = \frac{k - s_\pm}{k} O_1 + \frac{s_\pm}{k} O_2 = (x_\pm, y_\pm) = \frac{k - s_\pm}{k}(x_1, y_1) + \frac{s_\pm}{k}(x_2, y_2).$$

(2) If $0 < k < R_1 - R_2$ the circle $C_2[O_2, R_2]$ is inside the circle $C_1[O_1, R_1]$ (and do not intersect). The two Poncelet points are outside the segment O_1O_2 and closer to O_2. They are given by the same formula state in **(1)**. (Here, the solutions $s_\pm$ are different from the ones before but the formula for the Poncelet points is the same. Check it!)

(3) If $0 < k < R_2 - R_1$ the circle $C_1[O_1, R_1]$ is inside the circle $C_2[O_2, R_2]$ (and do not intersect). The two Poncelet points are outside the segment O_1O_2 and closer to O_1. Now, they are given by the formula

$$P_\pm = \frac{k + s_\pm}{k} O_1 - \frac{s_\pm}{k} O_2 = \text{etc.}$$

In **(1)**, **(2)**, and **(3)**, the foot of the radical axis of $C_1[O_1, R_1]$ and $C_2[O_2, R_2]$ is the midpoint of the Poncelet segment P_-P_+, that is, the point

$$F := \frac{1}{2}(P_- + P_+) := (f_1, f_2) = \text{etc.}$$

Then the **radical axis** of $C_1[O_1, R_1]$ and $C_2[O_2, R_2]$ is given analytically by the equation

$$(y_2 - y_1)(y - f_2) = -(x_2 - x_1)(x - f_1).$$

If the **circles are tangent**, then either $k = R_1 + R_2$, or $0 < k = R_1 - R_2$, or $0 < k = R_2 - R_1$. **Equation 3.8** has one (double) real solution, given by

$$s = \frac{k^2 + (R_1 + R_2)(R_1 - R_2)}{2k}.$$

Now, there is **one Poncelet point**, which is the point of tangency, call it $T = (t_1, t_2)$.

(4) If $k = R_1 + R_2$, then the circles are externally tangent, and $s = R_1$. The point of tangency is

$$T = (t_1, t_2) = \frac{R_2}{R_1 + R_2} O_1 + \frac{R_1}{R_1 + R_2} O_2 =$$

$$\frac{R_2}{R_1 + R_2}(x_1, y_1) + \frac{R_1}{R_1 + R_2}(x_2, y_2) = \left(\frac{R_2 x_1 + R_1 x_2}{R_1 + R_2}, \frac{R_2 y_1 + R_1 y_2}{R_1 + R_2} \right).$$

(5) If $0 < k = R_1 - R_2$, the circles are tangent and $C_2[O_2, R_2]$ is inside $C_1[O_1, R_1]$, and $s = R_1$. The point of tangency is given by

$$T = (t_1, t_2) = -\frac{R_2}{R_1 - R_2} O_1 + \frac{R_1}{R_1 - R_2} O_2 = \text{etc.}$$

(6) If $0 < k = R_2 - R_1$, the circles are tangent and $C_1[O_1, R_1]$ is inside $C_2[O_2, R_2]$, and $s = R_2$. The point of tangency is given by

$$T = (t_1, t_2) = \frac{R_2}{R_2 - R_1} O_1 - \frac{R_1}{R_2 - R_1} O_2 = \text{etc.}$$

In **(4)**, **(5)**, and **(6)**, the **radical axis** of $C_1[O_1, R_1]$ and $C_2[O_2, R_2]$ is the **common tangent**, and so it is given analytically by the equation

$$(y_2 - y_1)(y - t_2) = -(x_2 - x_1)(x - t_1).$$

(7) Next, the case left is when the **circles** $C_1[O_1, R_1]$ and $C_2[O_2, R_2]$ **intersect** (at two points). That is, when $0 \le |R_1 - R_2| < k < R_1 + R_2$. In this case, **Equation 3.8** has two complex conjugate solutions given by

$$u \pm i\,v = \frac{k^2 + (R_1 + R_2)(R_1 - R_2)}{2k} \pm$$
$$\pm i\frac{\sqrt{(k + R_1 + R_2)(k + R_1 - R_2)(k - R_1 + R_2)(-k + R_1 + R_2)}}{2k}.$$

If the real positive x-axis is $\overline{O_1O_2}$ with origin O_1, then the points of intersection are $A_\pm = (u, \pm v)$.

To find $A_\pm$, in general, it is convenient to complexify the points, use some complex arithmetic, and compute the rotation of $(u \pm i\,v)$ given by

$$w_\pm = \frac{[(x_2 - x_1) + i(y_2 - y_1)] \cdot (u \pm i\,v)}{k}.$$

Then,

$$A_\pm = [x_1 + \mathrm{Re}(w_\pm),\ y_1 + \mathrm{Im}(w_\pm)] := (a_\pm, b_\pm).$$

The **common cord is also the radical axis** of $C_1[O_1, R_1]$ and $C_2[O_2, R_2]$, and is given by

$$(b_+ - b_-)(y - b_-) = (a_+ - a_-)(x - a_-)$$

The **foot of the radical axis** on the center line is found by solving the linear system of this equation together with the equation of the center line O_1O_2, $(x_2 - x_1)(y - y_1) = (y_2 - y_1)(x - x_1)$.

Results of Section 4.5 Analytically.

The results of **Section 4.5** can be derived analytically. The real arithmetic computations are lengthy and so the synthetic solution that exposed there is efficient and convenient. However, the complex arithmetic is also convenient and we are going to use it to derive them.

Let us consider two inversions in complex-number form

$$f(z) = a + \frac{\rho}{\bar{z} - \bar{a}} \quad \text{and} \quad g(z) = b + \frac{\sigma}{\bar{z} - \bar{b}}.$$

Here, z is the complex variable, the centers of the inversions are the complex numbers a and b, and the **powers** $\rho > 0$ and $\sigma > 0$. So, the **radii** of the circles of the inversions are $\alpha = \sqrt{\rho}$ and $\beta = \sqrt{\sigma}$. We want to compute the composition $(g \circ f \circ g)(z)$ and draw the conclusions of **Section 4.5**.

First,

$$(f \circ g)(z) = a + \frac{\rho}{\bar{b} - \bar{a} + \frac{\sigma}{z-b}} = \ldots = \frac{\sigma a + [a(\bar{b} - \bar{a}) + \rho](z - b)}{(\bar{b} - \bar{a})(z - b) + \sigma}.$$

(This is a holomorphic fractional bilinear transformation, with determinant $\rho\sigma \neq 0$.) Then,

$$(g \circ f \circ g)(z) = b + \frac{\sigma}{\frac{\sigma\bar{a} + [\bar{a}(b-a)+\rho](\bar{z}-\bar{b})}{(b-a)(\bar{z}-\bar{b})+\sigma} - \bar{b}} = \ldots =$$

$$b + \frac{\sigma^2 + \sigma(b - a)(\bar{z} - \bar{b})}{\sigma(\bar{a} - \bar{b}) + (\rho - |b - a|^2)(\bar{z} - \bar{b})}.$$

At this point, we consider two special cases.

(1) $a = b$, i.e., the two circle are concentric. Then $\bar{a} = \bar{b}$ and

$$(g \circ f \circ g)(z) = b + \frac{\frac{\sigma^2}{\rho}}{\bar{z} - \bar{b}}.$$

This is an inversion with center $a = b$ and power $\dfrac{\sigma^2}{\rho} = \dfrac{\beta^4}{\alpha^2}$. This result agrees with **Figures 4.34** and **4.35**.

(2) $\rho - |b - a|^2 = 0 \iff |b - a| = \alpha (= \sqrt{\rho})$. That is, the length of the center segment is equal to the radius α. Then, the circles of the two inversions intersect and

$$(g \circ f \circ g)(z) = b + \frac{\sigma(a - b)}{|a - b|^2} - \frac{(a - b)^2}{|a - b|^2}(\bar{z} - \bar{b}).$$

This is the reflection in the line of the common cord of the two circles, in accordance with **Figure 4.30** and the calculation of the common cord in the previous subsection. (Work out the details.)

So, we now assume the generic case $\rho - |b - a|^2 \neq 0$. Then, we finally get

$$(g \circ f \circ g)(z) = \ldots = b + \frac{\sigma(b - a)}{\rho - |b - a|^2} + \frac{\frac{\rho\sigma^2}{(\rho-|b-a|^2)^2}}{\bar{z} - \left(\bar{b} + \frac{\sigma(\bar{b}-\bar{a})}{\rho-|b-a|^2}\right)}.$$

This is the inversion with center $b + \dfrac{\sigma(b - a)}{\rho - |b - a|^2}$ and power $\dfrac{\rho\sigma^2}{(\rho - |b - a|^2)^2}$. These results agree with the general results of **Section 4.5** and the remaining figures in that section.

Bibliography

[1] Baloglou, G. 2007. *Isometrica.* SUNY Oswego.

[2] Baxter, J., Fristedt B., Rogness J., 2017. *College Geometry I, Using Vectors and Calculus.* School of Mathematics, University of Minnesota.

[3] Coolidge, L. J., 2012. *A Treatise on the Circle and the Sphere*, classic reprint series. Forgotten Books.

[4] Coxeter, H. S. M., F. R. S. 1969. *Introduction to Geometry*, second edition. Wiley.

[5] Coxeter, H. S. M., F. R. S., Greitzer S.L. 1967. *Geometry Revisited.* The Mathematical Association of America, New mathematical Library.

[6] Dodge, C. W. 1972. *Euclidean Geometry and Transformations.* Dover.

[7] Frère, G.-M. 1926. *Exercises de Géométrie*, 7me, éd. Librairie Générale, Paris 1926.

[8] Greenberg, M. J. 1993. *Euclidean and Non-Euclidean Geometries, Development and History*, third edition. Freeman.

[9] Guggenheimer, H. W. 1967. *Plane Geometry and Its Groups.* Holden-Day.

[10] Isaacs, I. M. 2001. *Geometry for College Students.* Brooks / Cole.

[11] Johnson, R. A. 1929, republished 2007. *Advanced Euclidean Geometry.* Dover.

[12] Kakouris, M. 2008. *The Geometry of Inversion.* B.Science Thesis. University of Athens, Greece.

[13] Lemaire, G. 1926. *Méthodes de Résolution et de Discussion des Problèmes de Géométrie*, onzième édition. Paris Librairie Vuibert.

[14] Levrie, P. 2019. *A straightforward proof of Descartes' Circle Theorem.* The Mathematical Intelligencer, p. 24-27.

[15] Lowell, C. J. 2012. *A Treatise on the Circle and the Sphere.* Classic reprint series (1916). Forgotten Books.

[16] Ogilvy, C. S. 1969. *Excursions in Geometry.* Reprint. Dover.

[17] Pedoe, D. 1970. *Geometry a Comprehensive Course.* Dover.

[18] Posamentier, A. 2002. *Advanced Euclidean Geometry.* Key College Publishing.

[19] Roussos, I. M. 2022. *Basic Lessons on Isometries, Similarities and Inversions in the Euclidean Plane - A Synthetic Approach.* World Scientific.

[20] Roussos, I. M. 2010-2016. *Lecture Notes in Euclidean Geometry.*

[21] Roussos, I. M. 2012. "On the Steiner Minimizing Point and the Corresponding Algebraic System". The College Mathematics Journal Vol. 43, No. 4 (September 2012), pp. 305-308.

[22] Smart, R. J. 1993. *Modern Geometries*, fifth edition. Books / Cole.

[23] Yale, P. B. 1968. *Geometry and Symmetry*, Dover.

[24] Yaglom, I. M. 1973. *Geometric Transformations I, II, III*, translated from the Russian by Allen Shields and Abe Shenitzer. The MAA, New Mathematical Library.

[25] Yaglom, I. M. 2009. *Geometric Transformations IV, Circular Transformations*, translated from the Russian by Abe Shenitzer. The MAA, Anneli Lax New Mathematical Library, #44.

Index

www.ingramcontent.com/pod-product-compliance
Lightning Source LLC
Chambersburg PA
CBHW060832120126
37901CB00004B/38
* 9 7 8 9 8 1 9 8 2 0 1 5 3 *

www.ingramcontent.com/pod-product-compliance
Ingram Content Group UK Ltd.
Pitfield, Milton Keynes, MK11 3LW, UK
UKHW021014120926
8556IPUK00040B/1134

9 783631 764855